# SpringerBriefs in Applied Sciences and Technology

SpringerBriefs present concise summaries of cutting-edge research and practical applications across a wide spectrum of fields. Featuring compact volumes of 50 to 125 pages, the series covers a range of content from professional to academic.

Typical publications can be:

- A timely report of state-of-the art methods
- An introduction to or a manual for the application of mathematical or computer techniques
- A bridge between new research results, as published in journal articles
- A snapshot of a hot or emerging topic
- An in-depth case study
- A presentation of core concepts that students must understand in order to make independent contributions

SpringerBriefs are characterized by fast, global electronic dissemination, standard publishing contracts, standardized manuscript preparation and formatting guidelines, and expedited production schedules.

On the one hand, **SpringerBriefs in Applied Sciences and Technology** are devoted to the publication of fundamentals and applications within the different classical engineering disciplines as well as in interdisciplinary fields that recently emerged between these areas. On the other hand, as the boundary separating fundamental research and applied technology is more and more dissolving, this series is particularly open to trans-disciplinary topics between fundamental science and engineering.

Indexed by EI-Compendex, SCOPUS and Springerlink.

Jagat Shrestha

# Rural Road Development in Developing Countries

 Springer

Jagat Shrestha
Department of Civil Engineering
Pulchowk Campus, Institute of Engineering
Tribhuvan University
Lalitpur, Nepal

ISSN 2191-530X          ISSN 2191-5318  (electronic)
SpringerBriefs in Applied Sciences and Technology
ISBN 978-981-96-2011-1          ISBN 978-981-96-2012-8  (eBook)
https://doi.org/10.1007/978-981-96-2012-8

This Springer imprint is published by the registered company Springer Nature Singapore Pte Ltd.
The registered company address is: 152 Beach Road, #21-01/04 Gateway East, Singapore 189721,
Singapore

If disposing of this product, please recycle the paper.

# Preface

Effective road networks are essential in rural areas to foster local economic growth, which subsequently raises living standards for rural populations and enhances government service delivery. Although constructing roads in rural and mountainous regions poses challenges, these networks are vital for strengthening the local economy and improving quality of life. However, expanding road networks in such areas requires meticulous planning and implementation due to obstacles such as technical difficulties, management issues, and resource limitations, all of which can slow progress despite governmental efforts in developing countries.

Developing road networks in rural, particularly hilly, areas is a key intervention to increase physical access to rural settlements and public services, covering the majority of communities and facilities while making the best use of limited resources in developing nations. This book explores various rural road network concepts, both for new construction and upgrading. The "coverage-based rural road network" approach identifies strategic nodal points to connect settlements and public facilities within a region, thereby establishing a foundational rural network. Different models, including the typical rural road network structure and the backbone and branch (BB) network model as applied in Nepal's hilly regions, are examined.

The models in this book provide a portfolio of recommended links for road network enhancements and solutions tailored to different budget levels, optimizing transportation costs across a range of road surfaces (earthen, gravel, or asphalt) in rural areas. One specific model addresses a multi-objective approach to rural road network upgrades, balancing two primary goals: minimizing user operation costs and maximizing the population covered by road links within a set budget. The final model presents various efficient alternatives for decision-makers.

The models were tested in Nepal's Himalayan hill regions, demonstrating a practical approach to improving and expanding rural road networks in mountainous regions of developing countries. These models consider key regional characteristics—such as terrain, settlement distribution, public facility locations, and funding availability—without requiring extensive data.

This book serves as a guide for the planning, design, and construction of rural roads in hilly regions. Divided into six chapters, it is intended to support engineers, planners, and decision-makers involved in rural infrastructure development and public facility location planning.

I extend my sincere thanks to faculty members, practitioners, and all others who provided valuable suggestions, support, and encouragement in the completion of this book. My gratitude also goes to the authors whose research and practical experiences have been referenced throughout. While every effort has been made to ensure clarity and accuracy, some oversights or inaccuracies may remain; constructive feedback for future improvement is highly appreciated, and readers are encouraged to share their comments.

Kathmandu, Nepal                                          Jagat Shrestha, Ph.D.
August 2024

# Contents

# About the Author

**Dr. Jagat Shrestha** is currently Associate Professor at the Department of Civil Engineering, Pulchowk Campus, Institute of Engineering, Tribhuvan University, Kathmandu. He obtained his B. E. (Civil) and M.Sc. in Civil Engineering from Pulchowk Campus, Institute of Engineering, Tribhuvan University, and his Ph.D. from the University of Aveiro, Portugal. He has also obtained an MBA and an MA in Economics from Tribhuvan University. Dr. Shreshtha has over thirty years of experience in teaching, research, and practices in the fields of civil engineering infrastructure development, capacity building, and research management. His major areas of research interest include planning, development, and management of rural infrastructures, particularly rural roads, bridges, and buildings. He has published many research articles in respected international journals and also functions as Editorial Board Member of the Nepal Journal of Science and Technology.

# Chapter 1
# Introduction

## 1.1 Rational

Establishing essential infrastructure is not just an important endeavor; it is the cornerstone of societal progress and economic growth. Infrastructure fuels the expansion of businesses, sustains the daily functioning of communities, and drives the advancement of primary, secondary, and tertiary economic activities [1]. These systems spanning law enforcement, education, health care, water supply, transportation, and communication are indispensable for fostering equitable development [2, 3]. By meeting these basic needs, societies can unlock vast opportunities for improving the quality of life for their people. Moreover, infrastructure must reflect the unique demands and circumstances of urban and rural environments, ensuring that development is inclusive and sustainable.

Rural areas, often defined by smaller populations, lower population densities, and close ties to nature, face unique challenges. Agriculture dominates their economies, and their geographic characteristics, such as hilly terrains and isolation, often result in limited infrastructure development. Without adequate infrastructure, rural communities are left disconnected from national networks, depriving them of essential services and opportunities for economic and social growth. This disparity underscores the urgent need to prioritize rural infrastructure development.

The importance of rural infrastructure cannot be overstated. [4] identifies three key components of rural infrastructure: physical infrastructure (roads, water supply, electrification, and storage), social infrastructure (healthcare, education, and community services), and institutional infrastructure (financial institutions, agricultural support services, and cooperative frameworks). Each of these is a building block for development. Together, they create a foundation that enables rural populations to embrace innovation, engage with broader markets, and improve their standards of living. Without these critical elements, efforts to uplift rural communities remain fragmented and incomplete.

J. Shrestha, *Rural Road Development in Developing Countries*,
SpringerBriefs in Applied Sciences and Technology,
https://doi.org/10.1007/978-981-96-2012-8_1

Investments in infrastructure, as noted by [5], are multifaceted. From an economic perspective, infrastructure includes high-cost, capital-intensive projects like highways, water systems, and communication networks. From a governance perspective, it represents the public sector's tangible capital stock. Moreover, investments in research and development to build human capital represent a long-term commitment to sustainable growth. Infrastructure is not merely about building roads or installing utilities—it is about empowering communities, bridging divides, and creating environments where people can thrive.

Critically, infrastructure comprises both hard and soft elements. Hard infrastructure includes physical assets like roads, electricity, and irrigation systems, while soft infrastructure refers to institutional frameworks and corporate practices that provide services such as transportation, banking, and marketing. These two components are deeply interconnected. For instance, robust transportation systems (hard infrastructure) enable efficient market access, which, in turn, supports institutional growth in banking and trade (soft infrastructure). Neglecting either form hinders comprehensive development, emphasizing the need for balanced investments.

Scholars like [5] and [6] highlight the transformative potential of infrastructure investments in developing countries. Hard infrastructure such as roads, telecommunications, and energy systems directly boosts agricultural productivity and facilitates economic activities. At the same time, soft infrastructure healthcare, education, and social services empower individuals to lead healthier, more productive lives. Together, these elements form the backbone of rural and national development. Focusing solely on physical infrastructure, as some argue, is shortsighted; social infrastructure is equally vital for creating a well-rounded, resilient society.

Many developing countries face profound challenges in extending these benefits to their rural populations. A significant portion of rural communities remains disconnected from national networks, both physically and economically. This disconnection leads to lower standards of living and limited participation in broader social and economic activities. Poverty in rural areas is deeply linked to the lack of infrastructure, perpetuating cycles of deprivation and inequality. Addressing this issue requires a dual focus on physical connectivity and access to services such as education and healthcare.

For example, Nepal has several cases that illustrate these scenarios. In this largely rural country, the population is spread across mountainous and hilly regions, with many communities lacking access to even the most basic road networks. According to [7], only 30% of Nepal is accessible by road. In hilly regions, over 39% of the population must walk for hours to access essential services. This isolation is not just a matter of inconvenience; it is a barrier to economic participation, health care, and education, all of which are fundamental to improving quality of life. Extending and upgrading rural road networks are not just necessary; it is a matter of urgency. It is the key to integrating remote communities into the national framework, enabling access to services, and fostering economic resilience.

The barriers to rural infrastructure development are significant but not insurmountable. Financial constraints are among the most pressing issues, as rural projects often require substantial investments. However, the root of the problem often lies in inadequate planning and inefficient resource allocation. It is imperative to adopt innovative, integrated approaches that maximize the impact of available resources. Strategic planning can ensure that funds are directed toward projects with the greatest potential to uplift rural communities and connect them to larger markets.

Moreover, the development of rural roads plays a transformative role in addressing rural isolation and fostering economic opportunities. Roads serve as lifelines, connecting people to markets, schools, healthcare facilities, and social services. As [8] emphasize, rural road construction has a profound impact on improving living conditions. Similarly, [9] advocate for giving rural transportation systems the same priority as urban and intercity networks. Improved transportation infrastructure not only reduces travel times but also opens up access to goods, services, and economic opportunities, reducing poverty and empowering communities.

The benefits of infrastructure development extend far beyond economic gains. By improving access to healthcare, education, and other social services, infrastructure investments directly enhance the well-being of rural populations. This, in turn, creates a ripple effect, enabling communities to achieve self-sufficiency and sustain momentum in development activities [10]. It is not just about building roads or hospitals; it is about building futures, creating opportunities, and breaking the chains of poverty.

In conclusion, rural infrastructure development is a moral and economic imperative for developing countries. It holds the power to transform lives, empower communities, and fuel national growth. By prioritizing the development of transportation networks, public facilities, and essential services, governments and planners can create a more equitable and prosperous future. The challenges are significant, but so are the rewards. Infrastructure development is not merely an investment in physical assets; it is an investment in people, their potential, and their aspiration.

## 1.2 Objectives

The primary aim of this study is to explore a rural road design methodology that effectively incorporates connections to public facilities in rural areas. This overarching goal is broken down into the following specific objectives:

(a) Develop a planning technique for identifying optimal nodal locations within rural road networks.
(b) Formulate a methodology for defining the structure and layout of rural road networks.
(c) Create models that optimize rural road networks while ensuring integrated connections to public facility locations.

Although numerous studies have focused on infrastructure development, they often examine individual types of infrastructure in isolation, overlooking the complex relationships between them. For instance, the design of a transportation network can significantly impact the location and accessibility of public and private facilities. Even if facilities are strategically placed, a poorly designed network may fail to adequately serve residents. This underscores the need for a more integrated approach. As [11] and [12] emphasize, the interdependence between network design and facility locations makes it essential to address these factors simultaneously.

By integrating these considerations, decision-makers can make more informed and resource-efficient choices when working within financial constraints. For instance, when deciding whether to build schools, expand healthcare facilities, or improve road links, considering both network structure and facility placement can lead to better outcomes for rural communities [11].

This study underscores the importance of developing models for rural road networks that account for both existing and potential public facility locations. Such models can minimize costs, both in construction and operations, while maximizing the reach and functionality of the infrastructure. The primary focus is on enhancing rural road networks, particularly in hilly regions, to ensure that the majority of rural villages and public facilities are interconnected.

Through case studies, this research examines various network patterns and evaluates the applicability and effectiveness of the proposed models in real-world scenarios. These assessments consider both financial limitations and geographical challenges, providing a comprehensive view of practical implementation.

Moreover, the methodologies presented here have broader applications. Beyond road networks, they can be adapted for planning rural water distribution systems, electrification networks, and telecommunications infrastructure. By adopting such integrated approaches, rural infrastructure development can become more strategic, efficient, and impactful.

## 1.3   Outline of the Book

This book is structured into seven comprehensive chapters, each building on the last to provide a detailed exploration of rural road design and development methodologies, with a particular focus on hilly regions.

In this chapter serves as the introduction, offering a broad overview of the study's scope and significance. It outlines the objectives, general considerations, and structure of the book, setting the foundation for subsequent discussions.

Chapter 2 examines the application of various rural road models and techniques in developing countries worldwide. It highlights practical examples and explores the use of locally available materials that are frequently employed in rural road construction. This chapter underscores the adaptability of these techniques to different regional contexts.

Chapter 3 delves into existing rural road planning models in the literature, with a focus on the distinctive topographical features of hilly regions. While most conventional planning models are tailored for flat terrain, this chapter investigates their limitations and adapts core concepts to the unique challenges of hill roads. These insights serve as a precursor to the proposed rural road network model.

Chapter 4 shifts the focus to the specific context of Nepal, analyzing rural road construction in hilly regions. It explores local conditions, including the five-zone mountain model, to guide alignment selection and road construction strategies. This chapter bridges theoretical frameworks with real-world applications.

Chapter 5 introduces a covering-based rural road network model designed to address the shortcomings of existing approaches. In the context of hilly terrains, it proposes a novel planning framework, the covering-based rural road network model. The chapter considers the integration of public facilities such as health centers, schools, and rural markets into the network design. A significant finding is the identification of the backbone and branch (BB) network pattern, which is particularly suited to hilly areas. The methodologies presented can also be adapted to other rural infrastructure projects, including water supply systems, electrification networks, and telecommunications lines.

Chapter 6 explores rural road optimization techniques, presenting four distinct models for optimizing rural road networks:

Rural Road Network Model-1 (RRNM-1): Focused on improving rural road networks in both flat and hilly regions.
Rural Road Network Model-2 (RRNM-2): A general-purpose optimization model for creating new networks across different terrains.
Rural Road Network Model-3 (RRNM-3): Tailored to enhance existing links in hilly areas and optimize core networks in flat regions.
Rural Road Network Model-4 (RRNM-4): Geared toward designing new rural road networks in hilly areas and core networks in plains.

Models RRNM-1 and RRNM-3 are unique in their inclusion of three surface types: earthen, gravel, and asphalt. This chapter also evaluates the network models introduced in Chap. 5 and outlines four prioritization techniques for selecting rural road links for improvement. The compatibility of these models with prioritization approaches is analyzed in detail, ensuring practical applicability.

Chapter 7 presents a multi-objective analysis of rural road networks. It extends the optimization models discussed in Chap. 6, introducing a multi-objective rural road network model that incorporates the BB network structure for hilly regions. This chapter examines how decision-makers (DMs) interact with Pareto-optimal solutions and navigate trade-offs among competing objectives, providing actionable insights for infrastructure planners.

Each chapter builds on the last, combining theoretical insights with practical applications, to offer a holistic approach to rural road network planning, particularly in challenging terrains. This structured progression ensures that readers gain both the conceptual foundations and the practical tools needed to address real-world infrastructure challenges.

# References

1. A.O. Hirschman, *The strategy of Economic Development* (Yale University Press, New Haven, 1958)
2. A. Kahn, *Social Policy and Social Services*, 2nd edn. (Random House, New York, 1979)
3. A.L. Mabogunje, *Infrastructure in Planning Process* (University of London, Town and Country Planning Summer School England, 1974)
4. E.M. Gramlich, Infrastructure investment: a review essay. J. Econ. Lit. **32**(3), 1176–1196 (1994)
5. S. Wanmali, Y. Islam, Rural infrastructure and agriculture development in Southern Africa: a centre-periphery perspective. Geogr. J. **163**(3), 259–269 (1997)
6. R. Ahmed, C. Donovan, *Issues of infrastructure development: a synthesis of the literature Washington* (International Food Policy Research Institute, DC, 1992)
7. DoLIDAR Local Infrastructure Development Policy, (Ministry of Local Development, Government of Nepal, Kathmandu, 2004)
8. C. Gannon, Z. Liu, *Poverty and transport, TWU discussion papers, TWU-30* (World Bank, Washington, DC, 1997)
9. R. Tolley, B. Turton, *Transport Systems* (A Geographical Approach Longman Scientific & Technical, England, Policy and Planning, 1995)
10. J.B. Odoki, H.R. Kerali, F. Santorini, An integrated model for quantifying accessibility-benefits in developing countries. Transp. Res. Part A **35**, 601–623 (2001)
11. M.S. Daskin, S.H. Owen, Location models in transportation, in *Handbook of Transportation Science.* ed. by R.W. Hall (Kluwer Academic Publishers, Norwell, MA, 1999), pp.311–360
12. S. Melkote, M.S. Daskin, An integrated model of facility location and transportation network design. Transp. Res. Part A **35**, 515–538 (2001)

# Chapter 2
# Rural Road Models, Development Cases, and Construction Materials

## 2.1 Rural Road Models Practiced to Different Developing Countries

Different rural road models and approaches are suitable for different developing countries based on local conditions. The important activity is selecting and adapting the model that best fits the country's economic resources, environmental conditions, and sociopolitical context. Each approach has its strengths and can be tailored to meet the specific needs of the rural populations it aims to serve. The applicability of different models and approaches varies depending on a range of factors, including geography, climate, population density, economic resources, and governance structures.

The following sub-sections give an overview of how various rural road models and approaches applied in different developing countries:

### 2.1.1 Labor-Based Approaches

Labor-based approaches, which prioritize the use of local labor over heavy machinery, are particularly relevant in countries with high unemployment rates and abundant labor. This approach helps to create jobs, build local capacity, and reduce costs (ssatp.org). The Ethiopian Rural Travel and Transport Program used labor-based methods to construct and maintain rural roads, providing employment while reducing costs. The Rural Access Program in Nepal emphasized labor-based construction, engaging local communities in building and maintaining roads in hilly and remote areas.

© The Author(s), under exclusive license to Springer Nature Singapore Pte Ltd. 2025
J. Shrestha, *Rural Road Development in Developing Countries*,
SpringerBriefs in Applied Sciences and Technology,
https://doi.org/10.1007/978-981-96-2012-8_2

## *2.1.2  Community-Based Approaches*

Community-based approaches involve local communities in the planning, construction, and maintenance of rural roads. These are particularly effective in areas where community engagement is strong and where decentralized governance allows for local decision-making. The Rural Development Program in Bangladesh involved local communities in the maintenance of rural roads, improving sustainability and ensuring that the roads meet local needs, and the Community-Based Rural Development Project in Ghana [1] empowers local communities to take charge of road construction and maintenance, ensuring that roads serve the needs of the rural population.

## *2.1.3  Low-Cost and Gravel Roads*

Low-cost roads, often gravel or earth roads, are suitable for areas with low traffic volumes and limited financial resources. These roads are easier to construct and maintain but may require frequent maintenance in areas with heavy rainfall or challenging terrain. The Rural Roads Project in Tanzania [2] focused on constructing and maintaining low-cost gravel roads in rural areas, improving access to markets and services. In Kenya, low-cost roads have been prioritized in rural development strategies, particularly in arid and semi-arid regions where resources are limited.

## *2.1.4  Engineered and Paved Roads*

Engineered and paved roads are more expensive but offer greater durability and lower long-term maintenance costs. They are suitable for areas with higher traffic volumes, economic activities, and where weather conditions are harsh. The Pradhan Mantri Gram Sadak Yojana (PMGSY) program in India [3] focused on building all-weather, engineered roads to connect rural areas to the main road network, boosting economic growth and access to services. In Brazil, paved rural roads are common in agriculturally productive regions, where the investment in durable infrastructure is justified by the need to transport goods efficiently.

## *2.1.5  Integrated Rural Accessibility Planning (IRAP)*

IRAP is a holistic approach that integrates rural road planning with other aspects of rural development, such as access to markets, schools, and health care [4]. This method is particularly effective in countries prioritizing sustainable development and coordinating infrastructure planning with broader development goals. Vietnam has implemented IRAP to improve rural accessibility, linking road development with broader rural development plans to maximize impact. The Philippines uses IRAP to ensure that rural road development is aligned with national development goals, improving overall accessibility in rural regions.

### 2.1.6  Public–Private Partnerships (PPPs)

PPP models involve collaboration between government and private entities to finance, build, and maintain rural roads. This approach is viable in countries with a relatively stable economic environment and where the private sector is strong. PPPs have been used in South Africa to develop rural roads, leveraging private investment to improve infrastructure. Indonesia has employed PPPs in rural road development, especially in regions with high economic potential but limited public funding.

### 2.1.7  Climate-Resilient Road Models

In countries prone to extreme weather events, such as floods, landslides, or droughts, climate-resilient road models are crucial [5]. These models focus on building roads that can withstand harsh climatic conditions. The Climate-Resilient Rural Infrastructure Project in Bangladesh focuses on constructing roads that can endure frequent flooding. Sri Lanka's road development programs increasingly incorporate climate resilience to protect against monsoon impacts and rising sea levels.

### 2.1.8  Geographic Information System (GIS)-Based Planning

GIS-based planning is being used to identify optimal routes and design roads based on geographic, demographic, and economic data. This approach is especially advantageous in countries characterized by varied terrain and where it's crucial to allocate resources efficiently. For instance, Rwanda uses GIS to plan rural roads that optimize connectivity while minimizing environmental impact and construction costs. In Peru, GIS is used to plan rural road networks in the Andes, where challenging terrain requires careful planning.

## 2.2 Rural Road Development Cases for Different Countries

Rural road development has been a cornerstone of economic and social progress in many developing countries. The following sections demonstrate the diverse approaches to rural road development in different countries, each tailored to local conditions and challenges. Whether through labor-based construction, community involvement, integrated planning, or climate resilience, these programs have had a transformative impact on rural communities by improving connectivity, economic opportunities, and access to essential services. The success of these programs highlights the importance of context-specific strategies in rural road development.

Here are some specific cases that illustrate different approaches and outcomes in rural road development:

### 2.2.1 India: Pradhan Mantri Gram Sadak Yojana (PMGSY)

India has a vast rural population, with many villages lacking all-weather road connectivity. The lack of reliable road access was a significant barrier to economic development and access to essential services.

The Indian government launched the Pradhan Mantri Gram Sadak Yojana (PMGSY) in 2000. The program aimed to provide all-weather road connectivity to unconnected habitations in rural areas. It focused on building durable, paved roads that could withstand monsoon rains and other adverse weather conditions.

By 2023, PMGSY had connected over 178,000 habitations and constructed or upgraded over 700,000 km of rural roads. The program significantly improved access to markets, education, and health care, contributing to poverty reduction and economic growth in rural areas. The PMGSY also incorporated new technologies, such as the use of plastic waste in road construction, enhancing the sustainability of the roads.

### 2.2.2 Ethiopia: Ethiopian Rural Travel and Transport Program (ERTTP)

Ethiopia's rugged terrain and scattered rural population posed significant challenges for road development. Many rural areas were isolated, making it difficult for residents to access markets, health care, and education.

The Ethiopian Rural Travel and Transport Program (ERTTP) was launched in the late 1990s with support from the World Bank. The program emphasized labor-based construction techniques, using local labor to build low-cost, all-weather roads. The approach aimed to reduce construction costs, create local employment, and ensure community ownership of the roads.

The ERTTP improved access to essential services for millions of rural Ethiopians. By 2011, the program had built over 16,000 km of rural roads. The labor-based approach created significant employment opportunities in rural areas, boosting local incomes. The program also improved market access for farmers, leading to higher agricultural productivity and incomes.

### 2.2.3 Vietnam: Integrated Rural Accessibility Planning (IRAP)

Vietnam's rural areas, particularly in the mountainous northern and central regions, faced severe accessibility challenges due to difficult terrain and limited infrastructure.

Vietnam adopted the Integrated Rural Accessibility Planning (IRAP) approach in the 1990s. IRAP focuses on holistic planning that integrates road development with other rural development activities, such as improving access to markets, schools, and health care. The approach also prioritizes road investments based on economic and social impact.

IRAP led to significant improvements in rural accessibility, with thousands of kilometers of rural roads built or upgraded, particularly in remote and mountainous areas. The approach improved the targeting of investments, ensuring that roads were built where they would have the greatest impact on poverty reduction and economic development. IRAP contributed to Vietnam's overall rural development strategy, helping to reduce poverty and improve living standards in rural areas.

### 2.2.4 Peru: Rural Roads Program

Peru's diverse geography, including the Andes Mountains and Amazon Rainforest, made rural road development challenging. Many rural communities were isolated, with limited access to markets and services.

The Peruvian government, with support from the World Bank, launched the Rural Roads Program in the 1990s. The program focused on rehabilitating and maintaining rural roads using community-based methods. Local communities were involved in road maintenance, which helped to build local capacity and ensure the sustainability of the roads.

The program improved access to markets and services for over 3.5 million rural Peruvians by rehabilitating over 15,000 km of roads. Community involvement in road maintenance ensured that roads remained in good condition, reducing the need for costly repairs. The program also contributed to local economic development by improving market access and reducing transportation costs for rural producers.

## 2.2.5  Tanzania: Tanzania Rural Roads Program

Tanzania has a predominantly rural population, with many communities in remote areas lacking reliable road access. The country's rural road network was under-developed, leading to high transportation costs and limited access to markets and services.

The Tanzania Rural Roads Program was launched in the early 2000s, with support from various donors, including the World Bank and African Development Bank. The program focused on rehabilitating and upgrading rural roads to all-weather standards, using a mix of labor-based and mechanized approaches. The program also emphasized decentralization, with local governments playing a key role in road planning and maintenance.

The program improved access to markets, schools, and health care for millions of rural Tanzanians. By 2015, the program had rehabilitated over 10,000 km of rural roads. The decentralization approach empowered local governments and communities, improving the sustainability and relevance of road investments. The program contributed to increased agricultural productivity and rural incomes by reducing transportation costs and improving market access.

## 2.2.6  Bangladesh: Rural Development Program

Bangladesh's flat terrain and high population density make rural road development relatively straightforward, but frequent flooding poses significant challenges to maintaining road infrastructure.

The Rural Development Program, supported by various international donors, focused on constructing and maintaining rural roads that could withstand frequent flooding. The program used a community-based approach, involving local residents in road maintenance to ensure the sustainability of the infrastructure.

The program improved rural connectivity, reducing travel times and transportation costs for millions of rural residents. The community-based maintenance approach ensured that roads were regularly maintained, reducing the need for expensive repairs after floods. The program also contributed to rural economic development by improving access to markets and services, particularly in flood-prone areas.

## 2.2.7  Nepal: Rural Access Program (RAP)

Nepal's mountainous terrain and remote rural communities present significant challenges for road development. Many areas were isolated, with limited access to basic services and economic opportunities.

The Rural Access Program (RAP), funded by the UK's Department for International Development (DFID), focused on improving rural connectivity through labor-based road construction. The program aimed to create jobs, improve access to markets and services, and reduce poverty in remote rural areas. It also included a strong focus on environmental sustainability and gender equity.

RAP built or upgraded over 2000 km of rural roads, providing access to previously isolated communities. The program created significant employment opportunities, particularly for women and marginalized groups, helping to reduce poverty in target areas. RAP also implemented measures to mitigate environmental impacts, such as slope stabilization and drainage systems, ensuring the sustainability of the roads in Nepal's challenging terrain.

## 2.3 Construction Materials in Rural Road Construction

In rural road construction, particularly in developing countries, the use of local materials—those that are not traditionally considered high-quality or standard for road construction—has become increasingly common. These materials are often locally sourced and more cost-effective, which is crucial in resource-constrained environments.

The use of local materials in rural road construction is a practical and sustainable approach that leverages locally available resources to address the challenges of limited budgets and difficult terrain. These materials not only reduce costs but also contribute to environmental sustainability by recycling waste products and minimizing the need for new raw materials. The specific application of these materials varies depending on local availability, climate, and road requirements, demonstrating the importance of context-specific solutions in rural infrastructure development.

Here are some materials that are widely used in rural road construction around the globe:

### 2.3.1 Gravel and Natural Aggregates

Gravel and natural aggregates are some of the most common marginal materials used in rural roads. These materials are typically sourced from local quarries or riverbeds and are used for road bases and surfaces. Gravel roads are common in rural Kenya, where locally sourced aggregates provide a cost-effective solution for road construction. In many parts of rural India and Nepal, unpaved gravel roads are widely used due to the availability of natural aggregates, especially in regions where weather conditions are relatively mild.

## 2.3.2  Laterite

Laterite is a soil and rock type rich in iron and aluminum, commonly found in tropical regions. It is often used as a road-building material due to its strength and availability. Laterite is extensively used in Ghana for rural road construction, particularly in the coastal and forest regions where it is abundant. It provides a firm foundation and is relatively easy to compact. In the Amazon region of Brazil, laterite is used for constructing rural roads, taking advantage of its local availability and low cost.

## 2.3.3  Fly Ash

Fly ash is a byproduct of coal combustion in power plants. It can be used as a stabilizing agent in road construction, improving the strength and durability of subgrades and base courses. India uses fly ash in rural road construction under various government programs. It is mixed with soil or other materials to improve the structural integrity of roads, especially in regions with poor soil conditions. In South Africa, fly ash has been used to stabilize soils in road construction projects, providing an environmentally friendly solution while reducing costs.

## 2.3.4  Recycled Construction and Demolition Waste

Recycled materials from construction and demolition activities, such as crushed concrete, bricks, and asphalt, can be reused in road construction, particularly for base layers. In Malaysia, recycled concrete aggregate is used in the construction of rural roads, reducing the need for new raw materials and lowering construction costs [6]. The Netherlands has been a leader in using recycled construction materials in road construction, including rural roads, to promote sustainability and reduce environmental impact.

## 2.3.5  Plastic Waste

Plastic waste, particularly in the form of shredded or granulated plastic, has been increasingly used as an additive in road construction. It enhances the durability and flexibility of road surfaces, especially in hot climates. India has pioneered the use of plastic waste in rural road construction. The plastic is mixed with bitumen to create more durable and water-resistant roads, particularly in regions with heavy rainfall. Ghana has also experimented with using plastic waste in road construction to reduce environmental pollution and improve road quality [7].

### 2.3.6 Crushed Stone and Quarry Dust

Crushed stone and quarry dust are byproducts of the stone-crushing process and are used as fillers or as substitutes for sand in road construction. Quarry dust is widely used in Sri Lanka for rural road construction, especially in areas where sand is scarce. It provides a solid foundation for roads and is readily available near quarries. In Nigeria, crushed stone and quarry dust are used in rural road projects to improve the load-bearing capacity of the roads, particularly in regions with weak soils.

### 2.3.7 Rice Husk Ash

Rice husk ash, a byproduct of rice milling, is used as a pozzolanic material in road construction. It is particularly effective in stabilizing soils and improving the performance of road bases [8]. In Thailand, rice husk ash is used in the construction of rural roads, particularly in areas with abundant rice production. It helps to improve soil stability and reduce construction costs. Vietnam has also adopted the use of rice husk ash in rural road projects, leveraging its availability in rice-growing regions to enhance the durability of roads.

### 2.3.8 Coconut Coir

Coconut coir, derived from the husk of coconuts, is used as a reinforcing material in soil stabilization and erosion control in road construction. It is particularly useful in tropical regions where coconuts are abundant. In coastal regions of India, coconut coir is used to stabilize embankments and prevent soil erosion in rural road construction projects. Sri Lanka uses coconut coir in rural road projects to enhance soil stabilization and reduce the environmental impact of road construction.

### 2.3.9 Crushed Glass

Crushed glass is a recycled material used as an aggregate substitute in road construction. It is often mixed with other materials to improve road durability. In some rural areas of Australia, crushed glass is used in road construction as part of efforts to promote recycling and reduce landfill waste [9]. Certain rural road projects in the USA, particularly in states with strong recycling programs, have incorporated crushed glass into asphalt mixtures [10] to improve road longevity and sustainability.

## *2.3.10  Bamboo*

Bamboo, known for its strength and flexibility, is used in rural road construction as a reinforcing material in embankments and as part of composite materials. In rural China, bamboo is used in innovative ways, such as in composite materials for road construction and as reinforcement in soil stabilization projects [11]. In Indonesia, bamboo is employed in rural road construction to reinforce embankments and control erosion, taking advantage of its abundance and sustainability.

# References

1. Ghana—Community-Based Rural Development Project (English). Washington, D.C: World Bank Group. http://documents.worldbank.org/curated/en/954871468771003947/Ghana-Community-Based-Rural-Development-Project
2. The World Bank, Tanzania Roads to Inclusion and Socioeconomic Opportunities (RISE) Program (2020)
3. https://pmgsy.nic.in/
4. https://www.ilo.org/sites/default/files/wcmsp5/groups/public/@ed_emp/documents/publication/wcms_166079.pdf
5. https://inforse.org/asia/pdf/PUB_EVD_Eco-Village_Development_SouthAsia_2023-s.pdf
6. A. Milad et al., A review of the use of reclaimed asphalt pavement for road paving applications. J. Teknol. (Sci. Eng.) **82**(3), 35–44 (2020)
7. J. Appiah, K. Nkrumah, Use of waste plastic materials for road construction in Ghan. Case Stud. Constr. Mater. **6**, 1–7 (2016)
8. B. Kannur, H.S. Chore, Low-fines self-consolidating concrete using rice husk ash for road pavement: an environment-friendly and sustainable approach. Constr. Build. Mater. **365**, 130036 (2023)
9. https://www.arrb.com.au/latest-research/recycled-roads-a-glass-half-full
10. https://cptechcenter.org/ncc-projects/use-of-crushed-recycled-glass-in-the-construction-of-local-roadways/
11. L.C. Dlamini, S. Fakudze, G.G. Makombe, S. Muse, J. Zhu, Bamboo as a valuable resource and its utilization in historical and modern-day China. BioResources **17**(1), 1926–1938 (2022)

# Chapter 3
# Rural Road Planning Practices

## 3.1 Introduction

Policymakers are becoming more aware of the need of improving rural roads in developing countries. Some rural road planning approaches have been utilized for the methodical construction of rural roads. There is a substantial amount of literature on transportation network models, with a special emphasis on urban issues. As a result, there exist several innovative models for the construction and enhancement of urban transportation networks. Urban transportation network planning models primarily focus on selecting upgrades or additions to an existing network to address issues such as traffic congestion, energy consumption, and pollution management [1]. In contrast, rural transportation is centered on connecting and providing accessibility to rural and remote areas. Few studies deal with rural road network planning since it is still in its early stages.

Transport for rural areas often includes the movement of people and products to satisfy residential, economic, and social demands. This transportation can take place by various methods, including rails, routes, and roadways. In accordance with a study conducted by the World Bank and the International Labour Organization (ILO), transportation in rural regions is predominantly carried out on foot or with the assistance of intermediate means of transport (IMT) [2]. However, due to the lack of road access to the main network, these individuals are often cut off from the national network. As a result of the lack or low quality of physical access (transport) infrastructure, people have no or restricted access to products and services.

Rural villages are primarily dispersed in hilly areas owing to topographical factors. Due to local limitations and a lack of resources, linking all villages and utilities is difficult in such instances. As a result, accessibility to rural hill areas must be defined differently than accessibility in general. The current condition in the area is that villages and amenities can only be reached within a certain distance of a road. Consequently, accessibility in those areas needs to be understood in terms of

© The Author(s), under exclusive license to Springer Nature Singapore Pte Ltd. 2025
J. Shrestha, *Rural Road Development in Developing Countries*,
SpringerBriefs in Applied Sciences and Technology,
https://doi.org/10.1007/978-981-96-2012-8_3

coverage. This chapter emphasizes this issue and scrutinizes existing planning and development approaches and models in light of it.

The most fundamental approach to rural road design involves prioritizing settlements based on demographic and socioeconomic criteria and connecting them with the shortest road link. The majority of techniques rely on the Minimum Spanning Tree (MST) concept, taking into account factors like inter-settlement interaction, accessibility requirements, and so on. These methods are considered among the most effective and scientifically sound for rural road planning. While numerous studies, such as [3], view road network design as a subset of the Network Design Problem (NDP), only a few studies specifically focus on rural road networks developed for rural settlements.

The subsequent sections scrutinize existing methods and models addressing issues related to rural road network development. Briefly outlined are the principles and procedures relevant to this book. The shortcomings in current models and approaches are acknowledged within the context of rural areas, with a specific emphasis on hilly regions, and solutions are proposed to address these issues in the book.

## 3.2   Planning Methodology

This section provides a review of various existing methodologies for rural road planning and development. Examining these methodologies offers a foundational understanding to identify areas that may be lacking. These identified gaps can then be integrated into the further development of a methodology that is more contextually rational. The following subsections discuss the methodologies identified in the literature.

### 3.2.1   Priority Ranking (PR)

PR approaches ('Sufficiency Rating') were employed to plan highway maintenance and improvement in the early 1950s [4]. This method was among the initial ones employed to assess road linkages and has since been widely used in roadway planning and maintenance. Nevertheless, the approach has been adapted for rural road projects, as presented by Carnemark et al. (1976).

PR is a weighted rating technique where an overall rating score $S_i$ is determined for each proposed project $i$, represented by Eq. 3.1.

$$S_i = \sum_{j=1}^{m} W_j X_{ij},$$ (3.1)

where $W_j$ is the weight of the $j$-th considered factor or characteristic; $X_{ij}$ is the score of the $i$-th project for the $j$-th factor; $m$ is the number of factors. The higher the $S_i$ value, the more urgent is the project.

However, the process becomes challenging when project benefits are not independent, a common scenario in rural road design. In this situation, the advantage is derived from the connectivity of disconnected villages by a road and its connection with other roads, rather than from cost considerations.

### 3.2.2  Benefit/Cost Analysis (BCA)

Traditional road design strategies involve benefit/cost analysis (BCA). Carnemark et al. (1976) extended the BCA approach to rural road construction. Benefits are stated in monetary units and weighed against costs in BCA; the greater the benefit/cost ratio, the better the project. The fundamental challenge of BCA is accurately calculating all benefits in monetary terms.

The guiding principle in network planning for sparsely developed regions has been to enhance crop production cost reductions and meet access demands for farmers benefiting from network upgrades [5]. This idea is encapsulated in the "producers' surplus approach." Generally, resource-constrained countries opt for a widespread, low-quality network of small rural roads (earth or gravel-surfaced farm-to-market feeder roads and farm-access roads) over high-quality roads. This strategy reflects the concept of reducing road construction costs by providing low-quality roads while simultaneously increasing accessibility levels for rural areas, based on the advantages of rural road development. While this approach may be suitable for plain areas with high agricultural productivity, estimating the benefits of rural roads is challenging in many communities located in low-economic-potential (rural) regions.

[6] developed two computer-aided models in Nepal for planning and prioritizing district transportation networks in both developed and impoverished regions. GIS is used in the computer models to plan the district road network. The research used a combination of producer and consumer surpluses. In developed areas, the economic net present value (ENPV), economic internal rate of return (EIRR), and benefit–cost ratio (B/C ratio) are utilized to prioritize roads, while in undeveloped areas, socioeconomic criteria are considered rather than economic analysis. Notably, a significant portion of connections to rural communities lacks economic justification. Furthermore, the economic rationale for selecting rural road connectivity tends to avoid linking remote village communities, skewing the technique toward connecting densely inhabited and economically important areas.

BCA becomes more difficult in rural road planning when project benefits are not independent. Connection and accessibility are critical ideas in assessing rural road systems [7]. A set budget for road improvement is often assigned to a rural region in developing countries. As a result, budget constraints should be addressed while evaluating a rural road network. Typically, decision-makers are required to pick the optimal projects within the constraints of the available budget.

### 3.2.3   Centrality Index

The majority of trips in rural areas start in one population center and end in another. The centrality index, proposed by [8], can be utilized to assess the significance of towns identified as transportation nodes. Each settlement serves distinct purposes, functioning as service centers for areas such as education, health, business and commerce, industry institutions, and offices (e.g., bank, agriculture service center, veterinary office, post office, telephone office, electricity office, and cooperatives office). These functions attract traffic from neighboring towns and are consequently considered in the centrality index [9].

Each settlement's centrality index may be determined using Eq. 3.2 [9, 10].

$$C_j = \sum_{i=1}^{n} (W_i X_{ij}),$$
(3.2)

where

$C_j$   Centrality Index of the $j$th market center.
$W_i$   Weight of the $j$th marketing functions.
$X_{ij}$   Value of the $i$th function (number of establishments or shops at the $j$th market center).

Market centers are settlements that encompass marketing facilities, clinics, schools, and various other commercial, social, and welfare activities. The median threshold population approach may be used to calculate the weight of a function. The weight may be estimated using the procedure as shown in Eq. 3.3.

$$W_i = \frac{\text{Population  Median } i\text{th Function}}{\substack{\text{Lowest  median  population  of  the  market  centres} \\ \text{where  a  function  exists}}}.$$
(3.3)

The weight is calculated using the median threshold population approach as follows. The centrality index's anticipated growth can be based on past trends and surveys of aware people. In addition, an open-ended discussion about the evolution of market centers might be held with the informants [8]. The projection of urban expansion can be quite accurate given the historical trend of urbanization pattern and growing transportation networks. Nonetheless, government institutions do not always react quickly enough to market dynamics. Administrative office relocation decisions, for example, have severe political and social implications [6].

This index was also employed in [6] as a network module for district road network design and prioritization, aiding in the determination of the hierarchy of nodal points.

### 3.2.4 Intensity of Interaction

In a region, settlements interact with one another. The relevance of a link between two settlement/market centers may be determined by calculating the extent of interaction between them. Furthermore, the settlement interaction may be used to fix the settlement hierarchy. Aside from the people, the functions in a market center play an important role in generating or attracting visitors. Educational institutions, hospitals and private clinics, wholesale businesses, and other sectors should typically be included, as these functions attract traffic, as noted by [8]. To determine the interaction between two nodal points (settlements/centers), a gravity model can be employed. The centrality index is multiplied by the population. The centrality index may be thought of as the population's weight.

The distance between urban centers is crucial in generating trips based on a distance decay function. The force of interaction between two settlements may be calculated using the gravity model Eq. 3.4 [11].

$$I_{ij} = \frac{(W_i P_i)(W_j P_j)}{d_{ij}^b},$$  (3.4)

where

$I_{ij}$   Interaction between two nodal points $i$ and $j$.
$W_i$   Weight/centrality index of the node $i$.
$W_j$   Weight/centrality index of the node $j$.
$P_i$   Population of the node $i$.
$P_j$   Population of the node $j$.
$d$   Road distance between $i$ and $j$.
$b$   Exponent of $d$.

The $I_{ij}$ provides the preliminary indicative desire lines among the settlements. For the sake of simplicity the value of $b$ can be considered as 1 [8].

[6] named the transport demand estimating model after using Eq. 3.4 to calculate the level of interaction between two nodal points. However, in the study, the author specifies the value of $b$ as 2 [6]. The index includes education, health-related, commercial, and industrial services since they draw traffic from outlying areas. In terms of trip-producing capability, government institutions such as administrative offices, district-level courts, and police stations have little role compared to other tasks [6]. However, connecting to these services/institutions may be significant in terms of accessibility.

[12] generated, analyzed, and evaluated different rural road connectivity patterns using the concepts of settlement interaction, link efficiency, route efficiency, and network efficiency. To design the road network in such a way that it serves the study region in a balanced manner, an integrated area development method was proposed. The gravity hypothesis was utilized to measure inter-settlement interaction based on socioeconomic development, population, and settlement spatial separation. The

composite index of centrality was employed to assess the level of socioeconomic development. The interaction between two settlements was calculated as a function of the difference in their centrality scores. Alternative networks were constructed based on parameters such as maximum link efficiency, total link length, total operating cost, and completely built network. These were then evaluated in terms of overall cost (including construction and operational costs) to determine the best network. Despite its systematic consideration of numerous aspects of rural road planning, this approach has some weaknesses.

The Gravity hypothesis, employed to elucidate the underlying pattern of inter-settlement interaction, yields inaccurate results when the centrality scores of inter-acting settlements are identical, resulting in a zero interaction estimate by the model in such cases. The model's deterrent parameter was derived from the direct distance between settlements [12] and a basic connectivity matrix (Srivastava 1989), which does not hold true for new road linkages under hilly and irregular topographical conditions.

### 3.2.5   Road Density

Road networks constitute a grid in practice, and routes deviate to follow the optimal alignments for cost-effective and simple construction. Roads also tend to congregate in high agricultural output regions and market sites. Moreover, road density varies depending on agricultural access requirements. Furthermore, density in rural regions might vary based on topographical conditions.

The UNCHS [5] guideline employs the concept of road density, defined as the number of kilometers of road per square kilometer, to indicate the level of difficulty for a particular trip. The guidelines use this concept as a basis to determine the average distance that a crop on a farm or a person traveling to town must cover before reaching a road (to access any automated means of transportation). This concept is widely applied when designing the length of farm-to-market roads in local road networks. The average distance from a road to a farm in a network of parallel roads that are straight and uniformly dispersed in a region can be illustrated as shown in Table 3.1.

**Table 3.1** Road density and distance to roads [5]

| Density (km/sq. km) | Average distance to road (km) | Maximum distance to road (km) |
| --- | --- | --- |
| 0.500 | 0.50 | 1.0 |
| 0.200 | 1.25 | 2.5 |
| 0.100 | 2.50 | 5.0 |
| 0.050 | 5.00 | 10.0 |
| 0.025 | 10.00 | 20.0 |

### 3.2.6  Accessibility Indicators (AIs)

In developing countries, an Integrated Rural Accessibility Planning (IRAP) technique is also utilized to plan rural roads. The IRAP is a local-level planning technique that aims to optimize infrastructure investment based on the most pressing community requirements [13]. IRAP is based on the accessibility-activity approach and considers families' access demands as well as the activities that meet those needs. In its most basic form, IRAP tackles issues of home accessibility. It contributes to the rural accessibility planning process by assessing home access requirements as well as the infrastructure's ability to deliver access. These inputs can be used to develop strategies for reducing problems with access at various budget levels.

The IRAP is a process-oriented method capable of addressing traditionally considered transport-sector problems using either transport or non-transport techniques. For instance, if water collection is a crucial access demand, the issue can be addressed by improving trails or roads leading to the facility or by relocating water collection stations closer to the users. In this approach, the IRAP integrates mobility and service seating into the same framework.

The key features of the IRAP framework, as outlined by [13], include:

- It is needs-based, covering all aspects of household needs.
- It is comprehensive, offering solutions to access problems, not limited to transport issues.
- It is sustainable, as it is designed to be managed through local-level participation.

The IRAP methodology is founded on the access needs of households, as highlighted by [14] and [15]. The process commences with the collection of household data pertaining to service accessibility, including aspects such as water and firewood collection, healthcare, education, and so on. Various indicators are employed to identify the accessibility requirements for these services, with a primary focus on factors such as time or distance to the facilities providing these services. As outlined below, the IRAP methodology generates two principal outputs: accessibility indicators and accessibility profiles (Fig. 3.1).

The initial outcome of the household data gathering effort within the IRAP framework is the generation of Accessibility Indicators (AI) for each of the access needs. AI is defined by Eq. 3.5.

$$AI = \text{number of households} \times \text{time (or distance) to the facility.} \quad (3.5)$$

The number of households in the above equation represents the impacted population. The time (or distance) to the facility represents the amount of burden that the populace will bear. The greater the value of AI, the less accessible a certain facility will be to a given population. In this approach, the AI determines the inaccessibility of activities empirically. The AI can utilize parameters other than the number of households. Table 3.2 describes the range and significance of the AI parameters. It defines AI at various levels of rural accessibility planning.

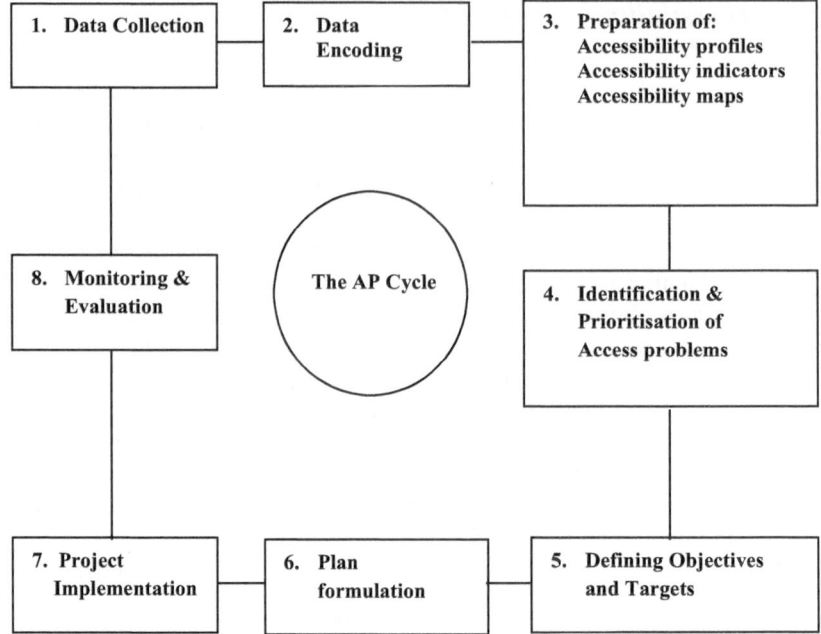

**Fig. 3.1**  Accessibility planning cycle [15]

**Table 3.2**  Examples of accessibility indicators (*AIs*) [14]

|  | Village | District | National |
|---|---|---|---|
| Water | Number of households × average collection time in the dry season | %age of households with no direct access to a water supply × average collection time in dry season | %age of households with no direct access to a water supply × average collection time in the dry season |
| Health | Number of households × time to a health center or clinic | %age of households in villages with no health center × average time to a health center | %age of households living in villages with no health center × Average time to a health center |
| Education | Number of primary school-age children × time taken to get to the school | %age of households with no primary school in their village × pupil/classroom ratio (or pupil/teacher ratio) | %age of households with no primary school in their village × pupil/classroom ratio (or pupil/teacher ratio) |

Table 3.2 indicates that the definition of AI is predicated on two factors:

(a)  The affected population; for example, households or the number of school-age children.
(b)  The burden, for example, the time taken to reach the facility or the distance to be covered.

In this manner, the AI outlines two potential approaches to addressing the accessibility problem:

- By reducing the size of the affected population, this can be achieved by enhancing the capacity of the facilities (increasing the number of classes, etc.).
- By reducing the distance or time for access, this can be accomplished by improving infrastructure (providing roads) or enhancing the supply of transport vehicles (IMT, NMT, etc.).

The AI is employed to generate accessibility profiles for the areas covered by the IRAP project. These profiles consist of maps of the entire study region that have been modified to depict the access difficulty of localities (e.g., villages) in relation to the burden they face. The accessibility profiles function as a planning tool, providing decision-makers with insights into the optimal utilization of resources.

The IRAP, the local-level planning method utilizes AI to prioritize rural infrastructure investments. It does so by examining rural household access to fundamental services and facilities such as health care, schools, markets, and water supplies. This approach is commonly employed for planning infrastructure improvements at the village level. It is important to note that rural roads are just one element of the planning process. This strategy is suitable for very low-traffic roads and village trails, even though it requires extensive data collection and is typically time-consuming. Nevertheless, this technology is currently being implemented in various projects in emerging Asian countries, including Cambodia, Laos, Thailand, the Philippines, Nepal, India, and Indonesia.

### 3.2.7  Costs

The costs associated with each link/network serve as the foundation for comparison and selection of the link/network to upgrade/select [16, 17]. Each link or network will have different levels of construction and travel costs. The optimal network is the one with the lowest overall cost, encompassing both construction and travel costs [16].

Additionally, [16] assumed that construction costs were proportional to link lengths. Similarly, travel costs have been calculated as a function of a quantity known as "person-kilometer (km)." The person-km for a village node is defined as the product of the population connected by the village node to its root node and the distance between that village node and its root node (root nodes are often settlements connected by a road or a point in the road). The following assumptions are included in the factor:

- The number of trips generated by a village node is proportional to its population.
- The travel costs are proportional to the distance traveled.

As a result, the person-km will be proportionate to overall travel costs.

Makarchi and Tillotson [17] separated costs into two groups in a rural road design model. One form of expense is known as construction costs, while the other is known as travel costs. Construction costs may be calculated rather accurately. However, it is challenging to calculate travel costs with a high level of certainty. In an area with relatively homogeneous terrain, it is reasonable to expect that construction costs will be proportional to connection lengths 17]. The cost of travel is anticipated to be proportional to (i) the number of persons linked by the link and (ii) the distance traveled through the link to reach the destination. According to [16], regardless of the specific nature of travel costs, they will be proportional to a factor called "person-km," defined as the product of the population connected by the link and the distance between the village and the destination via the link.

When the population of the communities and the linear distance between them are known, these approaches can be applied. This is particularly advantageous in rural areas where basic data can be readily accessible. Even when real costs are known, this strategy can still be effectively applied.

## 3.2.8   Accessibility Index

In developing countries, roads connecting rural villages may either be constructed from scratch or upgraded from existing paths, tracks, or fair-weather roads. All of these cannot be constructed/upgraded to the intended level due to budget constraints. As a result, the most efficient network that delivers the bare minimum of basic connectivity to all settlements must be chosen. The efficiency of road linkages in terms of accessibility levels needs to be evaluated. The accessibility provided by a road link is inversely related to the total travel distance required to fulfill the missing functions of a settlement. Total travel distance can be calculated by identifying several factors, including the missing functions, per capita trips for these functions, the location of functions in the region, the length of the connecting road link, and the settlement population. Total travel via the connecting road connection is the sum of person-km trips for different missing functions in the disconnected community. The less travel is required, the more accessible the settlement will be [18].

This approach can be employed to identify the road link that offers the highest accessibility for each disconnected community. The established network, based solely on person-km, seems to generate an optimal accessible connectivity pattern for disconnected settlements. To achieve this, settlements should be connected one at a time, analyzing and evaluating all available road link alternatives in a predetermined order. If this order is not followed, the possibility of interconnection among distinct disconnected communities may be overlooked. [18] introduces an accessibility indicator that considers settlements should be connected one at a time while analyzing and assessing all available road link choices in an orderly way. The accessibility indicator is calculated by dividing the total person-km traveled by the population of the disconnected settlement. The indication now shows the average person's journey to reach all of a disconnected settlement's missing functions via the connecting road link. It may be used to assess the accessibility provided by several connecting road link choices, and the one that provides maximum accessibility should be picked first in the network construction process. [18] developed the following accessibility index.

Road links are added to the existing network of road system in rural road design so that each disconnected community is connected to at least one road link. The connecting road link from the disconnected settlement connects to a neighboring connected settlement or any intermediary node on the current road network's linkages. As a result, the complete node system may be classified into two categories: unconnected nodes (settlements) and connected nodes (connected settlements and intermediate nodes on the current road network's linkages). A road link between disconnected and linked nodes is an example of a link option (Fig. 3.2). Unconnected settlements can be linked by converting existing trails, tracks, or unpaved roadways into connected settlements.

If the total number of linked nodes in the current road network is $m$, then there can potentially be $m$ connection choices providing access to each disconnected settlement. Many of these connectivity possibilities, however, will be redundant since they would intersect the current road network at numerous locations and will be too long in length. The connectivity option that provides the most access to the disconnected settlement should be picked. The accessibility of a link option can be calculated by identifying the unconnected settlement's missing functions (services) $k$, the total travel requirement of the unconnected settlement PK in terms of person-km to satisfy all of its missing functions, and the length of the new road link option $d$, as shown below.

The total person–km of travel, i.e., $\text{PK}i^{(l)}$, for an unconnected settlement $i$ to access missing functions $k$ through the link option $l$ can be calculated from Eq. (3.6).

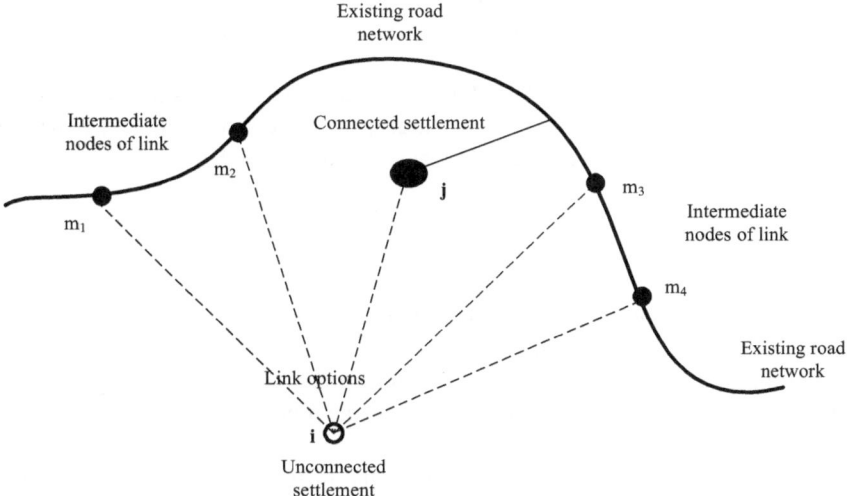

**Fig. 3.2** Link options from unconnected settlement [18]

$$PK_i^{(l)} = \sum_k T_i^k d_i^{k(l)}. \tag{3.6}$$

Here, $Ti^k$ is the total number of trips originating from unconnected settlement $i$ to access missing function $k$ and $di^{k(l)}$ is the distance between the unconnected settlements $i$ to the nearest function $k$ through the link option $l$.

The $di^{k(l)}$ can be calculated using Eq. (3.7).

$$d_i^{k(l)} = d_{im}^l + d_{mj}^k. \tag{3.7}$$

Here, $dim^l$ is the length of link option $l$ (i.e., the ground distance between the unconnected settlement $l$ and connected node $m$ on the network), and $dmj^k$ is the minimum distance between node $m$ and the nearest node $j$ on the network where the function $k$ is present. If the function $k$ is present at node $m$, then $dmj^k$ will be zero.

The accessibility index of link option $l$ (i.e. $Ai^l$) is calculated by dividing the total person-kilometer $PKi^{(l)}$ with the population of the settlement $Pi$ as given in Eq. (3.8).

$$A_i^l = \frac{PK_i^{(l)}}{P_i}. \tag{3.8}$$

Here, $Pi$ is the population of the unconnected settlement $i$.

It is assumed that the unconnected settlement access the first nearest function in its neighborhood, through the link option, to satisfy its missing functions. Higher value of $Ai^l$ indicates lower accessibility of link option $l$, and vice-versa.

In a physical sense, $Ai^l$ represents the average distance the population has to cross to access all the missing functions in the unconnected settlement when it gets connected through the road link $l$. The link option with minimum index value (i.e., maximum accessibility) is the most preferred choice in developing the maximum accessibility network.

The estimation of accessibility index, using Eq. (3.8), requires the determination of $Ti^k$ and $di^{k(l)}$. $Ti^k$ can be estimated by multiplying the settlement size $P$ with the appropriate trip rate $t^k$ for each function, as given in Eq. (3.9).

$$T_i^k = P_i t^k. \tag{3.9}$$

Here, the settlement size $Pi$ can be the settlement's $i$ population, number of households, etc. The trip rate $t^k$ represents trips per unit settlement size for missing function $k$ in the unconnected settlement.

Appropriate trip rate models for rural regions may be built by linking them with socioeconomic factors. Equation (3.10) depicts the model form.

$$t^k = f(\text{socio economic factors}). \tag{3.10}$$

In the absence of any such model, $t^k$ can be estimated for a few of the functions. For example, if the whole rural population is to be educated to the high school level by the end of the planning horizon, the trip rates to primary, middle, and high school will be proportional to the proportion of the school-going population in their respective age groups. The trip rates for marketing, health, and other purposes, on the other hand, will be determined by the area's predicted degree of growth. These trip rates will be lower in developing areas and higher in developed areas.

The term $di^{k(l)}$ has two components, i.e., $dmj^k$ and $dim^l$. Here, the term $dmj^k$ can be estimated quite easily, as the location of facilities on the existing road network is known. The other term $dim^l$, which is the length of new link option, will depend on its actual alignment on the ground. This alignment does not necessarily have to be direct and will be determined mostly by the topography of the area. Water bodies, hills, expensive constructions, terrain type, and other topographic characteristics may induce variation from straight alignment. Even with a small number of disconnected communities, there will be several road link possibilities and alignments depending on the terrain of the area. These connection choices must be evaluated for alignment, length, construction cost, and other factors in order to pick the least expensive option [18]. Furthermore, huge quantities of data are required for numerous services in order to estimate the accessibility index of each link.

## 3.2.9 Facility-Based Approach

Apart from market centers, a lot of rural areas might need additional access to other services like banking, administrative offices, health care, and education. In addition to market centers that integrate access to these facilities, [19] developed a facility-based model to assess accessibility to these facilities. The model considers how close the village is to market centers and specific institutions for user-friendly amenities. The local market or the closest educational facilities are the destinations of about 93% of rural travels. According to [19], approximately 93% of the travel needs of the rural population may be satisfied if road connections are made to the closest market and educational centers.

## 3.2.10 Traffic Flow

The foundation of the traditional transportation planning method is the division of a study area into smaller zones and sub-zones, followed by an Origin–Destination (O-D) matrix analysis of inter-zonal trip frequencies to ascertain the necessary connections between different origins and destinations. Rural road design may not be a good fit for this approach. There will be roughly 1500–2000 rural villages even while at the district level, so gathering trip frequency data for each settlement to create the O-D matrix will require a significant amount of resources. The observed trip frequencies will remain comparatively low regardless of whether the O-D pattern is made available, which would hardly justify the links in a network. The purpose of this data is solely to ascertain the links' relative priority [12]. This proves the impracticability of traditional methods.

In addition, it can be difficult to project how much traffic a rural road will handle when bridge construction and significant improvements are taken into account. In rural areas, where traffic data is frequently elusive, the population is considered to be a good stand-in for traffic [19].

The majority of research on transportation infrastructure planning focus on selecting upgrades or additions to an already-existing network to reduce pollution, energy use, traffic congestion, or other relevant factors. Developing countries face different network design challenges than developed countries when it comes to rural road networks. A small number of rural roads may still exist, but the networks are being constructed around the current main highways. Moreover, traffic congestion may be regarded as having little effect on travel time in rural areas of developing countries due to low-traffic volumes. According to Makarchi and Tilloston 17], the specific objective is to link every settlement—regardless of size—to the network.

[20] examined various rural roadway investment strategies in the USA as well as the design of rural road networks using a traffic simulation model. A technique for managing rural roads economically was found by examining both probabilistic and deterministic traffic simulation models. The authors have trouble determining trip generation from rural communities and getting traffic data from remote settlements. In rural areas, it is usually more difficult to obtain traffic data and analysis, and the results are also less important. To get around this, rural road planning models use person-km [16–18] or population [19].

## 3.3   Rural Road Network Generation

One of the first studies in the field of rural road network formation was done by [21], who created a system methodology for rural road development using the idea of graph theory (Fig. 3.3). Markets and main routes were perceived as high-intensity electrical charges that attracted lower charges from adjacent communities. In this model, the villages were connected to the closest market or to any nearby highways by means of the MST. The weight of multiple alternate road connections for a village was estimated by applying the gravity model to build a complete road network design that links all villages to nearby market centers by selecting linkages with the highest weight.

Later, UNCHS [5] guidelines point to two major classes of network models that can be used for network construction: (i) the "floating points" system, in which road junctions in the network are allowed to occur at any location in the plane, and (ii) the "fixed point" system, in which junctions are confined to a finite set of locations (e.g., cities, market towns, and villages). Models dealing with spanning tree problems are very useful in this scenario. The implementation of this principle is shown by a model

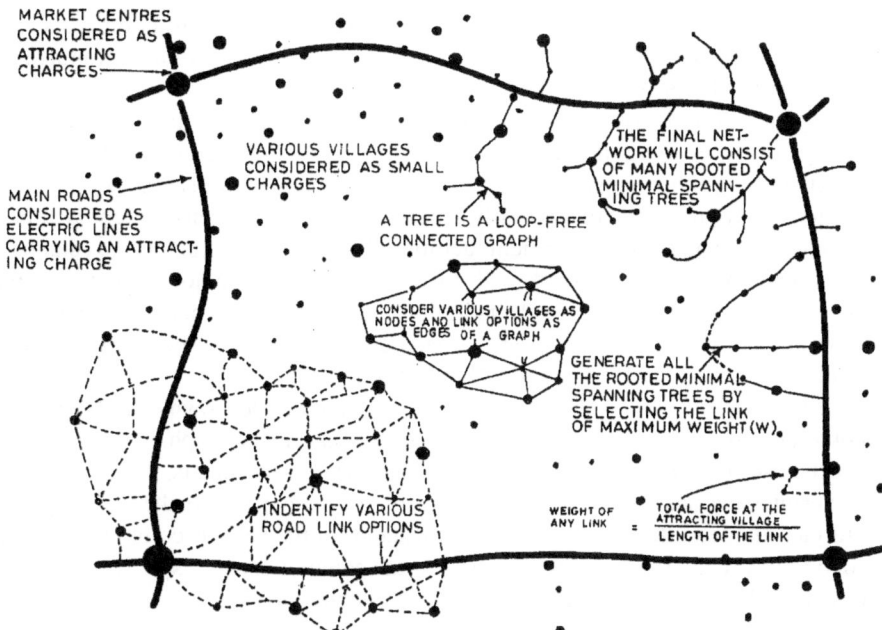

**Fig. 3.3** Approach to rural road development [19]

devised by [22], which aims to optimize overall flow in the network (calculated using gravity models) within the restrictions of a particular road mile.

UNCHS [5] proposed procedures for generating network connection options. Options are restricted to a handful that is evaluated for comprehensive engineering/ economic evaluation based on a set of simple evaluation criteria or indicators. It also recommends evaluating the road network through the development of indicators based on graph-theoretic measures of network connectedness and/or gravity concept-based measures of geographical accessibility. According to the recommendations, the disadvantage of these measurements is that they are restricted to the investigation of topological aspects of the network and do not address variables such as capacity, kind of usage, and construction cost. The technique, on the other hand, seeks to determine the most accessible paths connecting settlements.

A method for planning rural road networks that minimizes overall costs, which includes construction and trip expenses, was proposed by [16]. In this process, rural settlements are connected with different link possibilities to generate alternative networks from a set of specified roadway linkages. The optimal network was chosen as the one with the lowest cost of construction, as determined by the MST. The model develops rural road network designs by offering the most cost-effective all-weather road link from each hamlet to neighboring main roads or market centers. The most optimal rural road network is developed by minimizing overall transportation costs between a predefined set of villages, market centers, and main roadways in order to provide road links to villages. The approach can also reduce construction costs at a given degree of network performance. The model considers the overall advantage to be the accessibility provided to the settlements.

Trip generation, distribution, and assignment are not as important in rural road planning as they are in traditional transportation planning methodologies. The necessary statistics are either unavailable or impractical to obtain in rural areas. The system presented by [16] uses data that is easily accessible from public census records, topographical maps, and local government facilities. The aim of this model is to establish a fundamental rural road network with road connectivity to major market centers, medical facilities, and educational institutions, as well as commercial, social, and welfare operations in each settlement in road network planning. The market centers are said to attract traffic from neighboring communities.

This approach uses a "graph" model that looks at numerous towns, marketplaces, major roadways, and rural roads. Nodes and edges make up the graph. Road links that connect these nodes are referred to as edges, and various settlements, market centers, and road intersections on significant roads are referred to as nodes. In the model, nodes are further divided into village nodes, which stand for the various communities that need to be linked, and root nodes, which show the potential locations to which village nodes could be connected. When the network is complete, it looks like a group of rooted trees.

The construction and transport cost will vary depending on the character of the road networks. The distances that must be covered to reach root nodes may grow extraordinarily long, increasing travel expenses, if construction costs are reduced (by selecting linkages of the shortest possible length). The cost of construction will increase if, on the other hand, the links that have the shortest lengths to the root nodes are selected because each node will have a nearly direct connection to it. The network with the lowest total cost (construction + trip costs) is the ideal network connection. Although different networks will result in different trip patterns, the model actually simplifies by employing a fixed trip matrix.

The model constructs a minimal construction cost network initially, then iteratively decreases the overall cost to a minimum. As a result, the network achieves optimum performance. The model employs a conversion factor: "RC" to normalize construction and transport costs (which are measured in kilometers and person-km, respectively). The importance of this statistic is that one unit of connection length costs the same as one kilometer of travel to RC individuals throughout the course of the project. In other words, an extra kilometer of link length is required in order to save one kilometer of RC transit time to the root node. The value of the RC determines the optimal network that can be established.

For developing countries, [7] proposed a methodology for planning roads in rural areas. The study describes a combinatorial tree design problem as the primary decision model, evaluates several approaches to the problem, and offers both exact and heuristic solutions. It also suggests a novel network modeling methodology. Many families in developing countries often lack sufficient access to local market centers and essential retail and service businesses, including small industrial operations situated near the local market. Many hamlets and small settlements become entirely isolated after intense tropical rains for weeks or months. Some are only reachable by road vehicles, requiring significant diversions. Circular routes are the only accessible linkages with the regional exchange and distribution system at certain times of year. Prioritizing the construction of new all-weather roads is essential, aside from the necessary upkeep of the current networks of earth roads. These all-weather roads should link as many villages and hamlets as possible to the nearby market. The actual travel times on the network of all-weather roads can be overlooked to some extent. Those who live in rural areas are used to traveling a greater distance during the rainy season. It is important to utilize the available, but never enough, finances to prevent seasonal inaccessibility in as many communities as feasible. Travel time considerations must be subordinated until all villages can access all-weather roads, as this kind of isolation is highly unfavorable. Optimizing the number of villages linked to the nearby market through all-weather roads is, thus, a suitable quantitative objective for the creation of effective rural road networks. Given that hamlets and villages fluctuate in size, it is best to consider the various population concentrations and optimize the most optimal number of dwellings to connect. Household numbers are typically public knowledge.

According to [7], factors other than population are typically less significant in distinguishing rural settlements. Usually, there are no stand-out facilities or activities in these locations that could draw more traffic. Since these roads usually offer sufficient access to the local market(s), connecting such villages to a provincial route is often sufficient. Rural roads in developing countries are rarely new. Most road projects provide improvements to already-existing rails or roadways. An initial network N comprising all of the existing rural road segments and any prospective new segments was used to create the rural road decision model. The original network's vertex V represents population concentrations, or "villages." No population is connected to edges E. The model's only difficulty is connectivity, ideally, carefully chosen new roadways combined with pertinent current segments would create a tree T within network N. According to [7], this issue is referred to as the Weighted Subtree Problem (WSP). This can be resolved as a mixed integer linear programming issue.

WSP can be viewed as a subset of NDPs and seems to be a workable decision model for rural road planning in developing countries [3]. If funding were available to link all settlements, these issues would take on greater significance. From the initial graph, N (V, E), decision-makers select a spanning tree, but there is not enough funding for its development. In line with the district's funding allotment, the tree network N must then be implemented gradually.

To increase the effectiveness of the resources readily accessible for rural roadways, [19] devised a computer-based, user-friendly method for systematic planning. A cheap, all-weather road link between every community and the nearby market and educational centers are provided under the plan. It uses a heuristic approach to lower the overall transit cost of the network.

In general, two basic methodologies are utilized to prioritize rural roads: (a) sufficiency assessment and (b) cost–benefit analysis. The cost–benefit analysis entails evaluating multiple costs and advantages connected with a road in the same monetary terms, which is a challenging undertaking. The approach prioritizes rural roads based on a basic criterion, the population serviced per unit investment. As previously said, accessibility is seen as the primary advantage of rural road investment. Thus, the population serviced by a link divided by the cost of construction may be used as a decent proxy for the estimated benefit of a rural road link. The link servicing the most people per unit investment is given top priority. A rural road link's priority can be computed according to Eq. 3.11.

$$\text{Priority for a road link} = \frac{\text{Population served by the link}}{\text{construction cost of the link}}. \tag{3.11}$$

In addition, a facility-based network development mechanism has been presented in the same model, which will be explained below.

Every village that is to be connected is regarded as an unconnected node. Villages located on main highways or at the intersection of rural road linkages with main roads are termed linked nodes. The road linkages that emerge from the linked nodes and may be utilized to connect any disconnected hamlet are identified, and their building costs are evaluated. Among the numerous linkages coming from the linked nodes, the one with the lowest construction cost is picked. The selected link is evaluated based on the criterion of distance to a service center. In the event that the link fails to fulfill the necessary criteria, more links are selected (in increasing order of construction cost) until the link that satisfies the desired criterion at the lowest feasible cost is selected. Additionally, the village node that is connected by the chosen link is included among the connected nodes.

However, some villages might not be connected since there aren't enough services within the allowable specified distance. In this case, either more facilities must be constructed to meet the suggested standards or the maximum distance from a particular facility must be raised. The procedure is completed when every unconnected village node is connected to the root node, either directly or via other village nodes.

[8] created settlement-based interaction models for the creation of rural road networks (Fig. 3.4), utilizing the centrality index and gravity model in conjunction with the pre-existing rural transportation infrastructures as a foundation.

The initial steps involve identifying the nodes and the existing road network. The existing road network connecting the specified nodes can be located. It is possible to compute a shortest route matrix connecting different nodes in the network by utilizing the road network and nodal positions. The missing links, which show which roads, are candidates for building or repair may be found after calculating the shortest pathways and mapping the existing road network.

The above-described processes are used in developed areas in their basic form. Distinct strategies are employed for developed and underdeveloped regions. Still, most planning initiatives are similar to those in developed areas [6, 23]. In general, underdeveloped regions have no or extremely few rural road connections. As a result, in undeveloped areas, the primary problem is to ensure accessibility within a suitable time or distance.

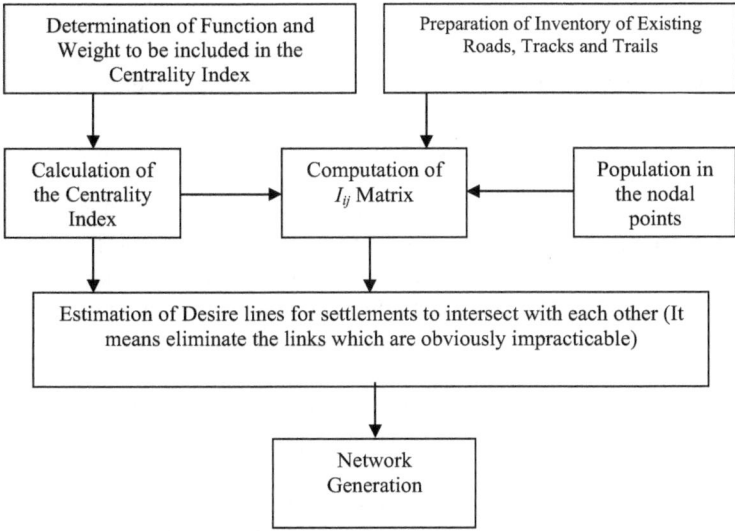

**Fig. 3.4**  Settlement-based network generation model [8]

In order to create district road networks in undeveloped areas, accessible regions—connected by country roads—are divided from inaccessible areas according to the interaction model [8]. The gateway and hinterland nodal points are established after distinguishing between accessible and inaccessible locations. The process of identifying nodal points is similar to that of developing regions. In the second stage, the closest hinterland nodal points are connected to the gateway nodal points. The shortest path problem might not be as significant in deficient areas since there might not be any alternative options. Furthermore, selecting the optimum alignment is crucial from an engineering perspective.

[18] proposed and used GIS technology to implement a plan for creating rural roadways based on accessibility considerations. A new accessibility index is developed, which assigns a value to different link options according to how well they enable access to the functionalities that are absent from the disconnected settlement. By keeping an integrated road system, this accessibility-based approach to rural road construction also offers unconnected settlements the best value in terms of coordinated access to numerous facilities or the main road network. More rural road connections to already-existing routes are prioritized. The connecting alternatives include either a fully new road alignment or the area's existing trails and tracks.

For every isolated community, the accessibility model assesses all possible connections (such as linking to a nearby road or an already-connected settlement) and calculates each community's accessibility index. For every unconnected settlement, the link option with the highest accessibility (i.e., lowest index value) is determined. This procedure is done for each unconnected settlement. Following that, the accessibility index values of all the link possibilities of distinct unconnected settlements are compared (explained in 3.2.8). The link option associated with the unconnected settlement that has the lowest accessibility index value is identified. The existing network can be expanded with the addition of the first-priority rural road link.

At this point, the list of linked and disconnected villages, as well as the current road network, is updated. By revising the distances from the facilities to the connected settlements, the newly connected community now offers its services to additional communities nearby that can be reached via the updated network.

At this time, there are new options for connecting the recently linked community to the other unconnected settlements in the vicinity; these should have their accessibility evaluated. The link possibilities with the lowest index value (i.e., highest accessibility) are found for the remaining disconnected communities. These new connection options' accessibility index is computed and compared with the accessibility values of the link options for the disconnected settlements. The new link choices and index values are applied to those disconnected settlements with lower accessibility index values. Again, the road network is expanded to include the lowest index value link choice and the settlement that goes along with it. This process is repeated until every community is connected. Figure 3.5 depicts the flow chart of the approach described above for developing rural road networks based on accessibility criteria.

They overlap whenever the produced road linkages intersect with existing country roads. It can be seen that all of the formerly disconnected communities were linked by a single road link, providing optimum accessibility to the settlement. It is possible to identify road portions in the current rural road network that are not required to access the functions that isolated settlements lack. Furthermore, by changing the connection selection criteria, it is possible to construct different types of rural road networks. Construction costs may also be used as a criterion for generating the least expensive rural road network.

Using topographic data from the area, [18] has also developed a novel method that employs GIS to determine the least expensive road configuration between any two locations.

## 3.4   Solution Techniques

To tackle the rural road network problem, numerous authors utilized various methods and solution strategies. The precise method for establishing link choices between village nodes and root nodes was presented by [16]. A partially created tree's root nodes are connected nodes that will eventually be attached to unconnected village nodes. We select and join the detached node that is closest to the connected nodes. Therefore, the village node is thought of as a connected node that can be joined to other unconnected nodes. The procedure is repeated to create links between the remaining unconnected nodes. The resulting tree is the shortest network to construct, needing just the shortest distances between village nodes and root nodes.

According to [16], the generation of a rural road network for a given value of RC is as follows:

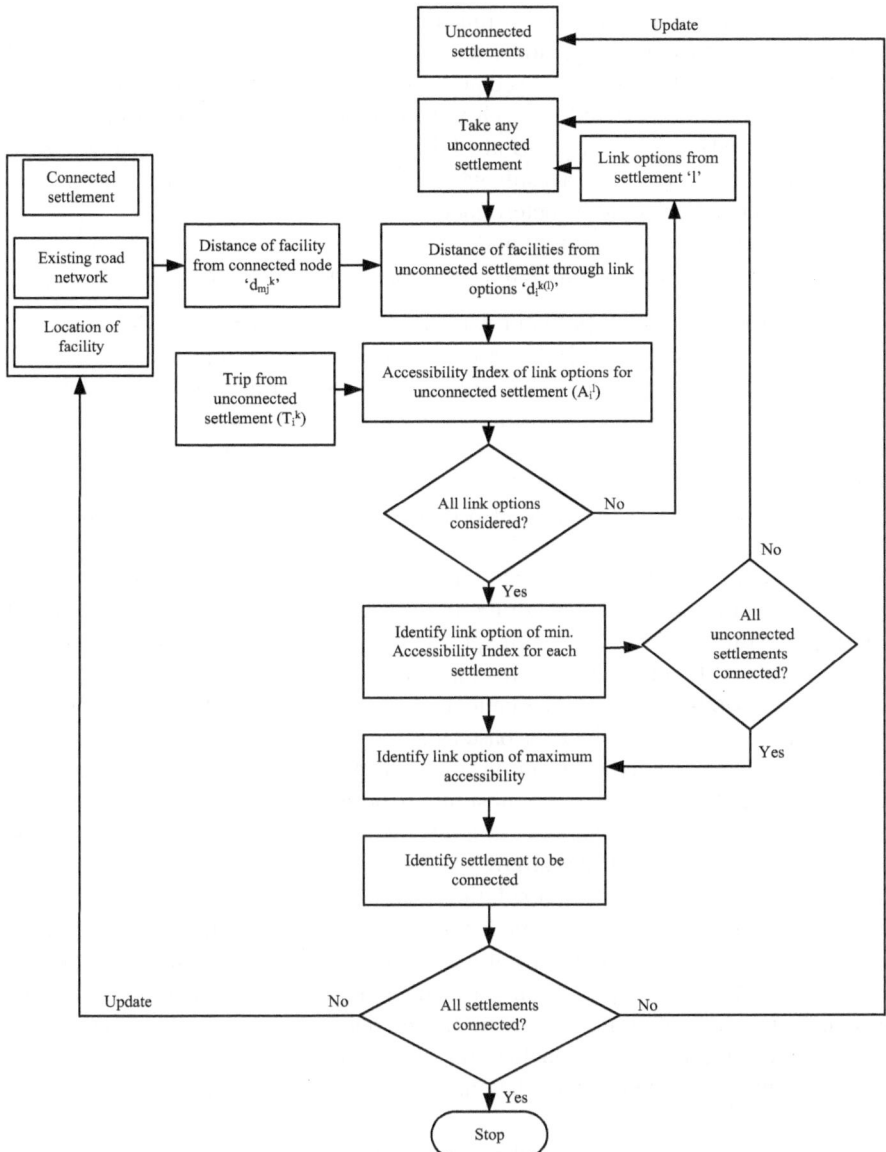

**Fig. 3.5**  Rural road network generation based on accessibility criteria [17]

1. Determine the distance between a village node and its root node (connected), as
   well as the population linked by that village node to its root node (the sum of
   these two will yield the person-km for that village node).

2. Begin with a village node that is linked to its root node. It might also be linked to other prospective village nodes and linkages. Calculate the total cost of each alternate connection for the linked node and choose the one with the lowest cost.
3. The population connected by each village node and its distance from the root node are used to pick linkages. Both of these characteristics change if any existing network link is altered. As a result, whenever a connection is changed, the person-km of each node should be determined.
4. Repeat steps (2) and (3) for all nodes.
5. Repeat steps (2) to (4) until there is no change in the network.

The rural road network problem was employed as a WSP in a different rural road network problem by [7]. Accurate resolution requires the use of a specific purpose enumerative approach. The method is not appropriate for rural road design, even while it yields optimal solutions for a wide range of WSPs and can be used as a heuristic for much larger issues. The framework for rural planning demands a straightforward solution technique that is both manual and easily programmable on a microcomputer. Planners should also find the technique appealing. It should be built on ideas that people can readily grasp and are familiar with. It is possible that an exact, optimum solution for WSP is not necessary. A minimal number of near-optimal network configurations are required, which can be utilized as benchmark solutions for further consideration. It appears logical that a basic heuristic method is a better solution strategy than one that guarantees optimality.

For WSP, a well-known and easy heuristic for the traditional 0-1 Knapsack Problem (KP) [24] is employed, which is based on the ranking of utility/cost ratios for each project (Continuous KP solution method). Additions to projects increase the budget ceiling. Subsequently, by substituting projects other than the solution for those already included, the solution might be enhanced.

Two strategies were put forward by [7]. The first is to use a quantitative network optimization model to find good solutions. To assess and narrow down the available possibilities, this model often includes (approximate) utility and cost data. Secondly, it is possible to modify the given solutions by incorporating minor qualitative elements. The various possibilities are frequently contrasted. An in-depth benefit/cost analysis or priority ranking may be carried out at this stage if there is sufficient time and expertise. Using these tactics could help narrow down the options to a select few, thoroughly considered, route layouts. Developing countries frequently engage in this practice.

Two techniques for creating rural road networks were developed by Makarchi and Tilloston (1991). The first algorithm chooses exactly one connection from among the available link possibilities for every disconnected node to determine the MST network. It was referred to as the Shortest-Spanning Tree (SST). The shortest total length of the road, or the MST network, is the one that links all of the communities for the least amount of money. In this situation, the additional construction is so minor that nearly any savings in trip expenses would justify this alteration to the established network. The method for building the MST network in this case splits the nodes into two groups: the connected set, which initially consists only of the root nodes, and the unconnected set.

Figures 3.6 and 3.7 depict the graph of a typical area and the matching MST for the typical area.

Using the connection options, the disconnected set node that is nearest to any member of the linked set is found, removed from the disconnected set, and added to the linked set. The link involved is also documented. The set of connections used defines the MST, and the process repeats until every node is in the connected set.

Since the units used to represent these expenses are not exactly equivalent—kilometers and person-kilometers—a conversion factor Lc is developed and assigned to the link lengths as Eq. 3.12.

$$\text{Total costs} = \text{person} - \text{kilometers} + (\text{Lc} \times \text{kilometers}). \qquad (3.12)$$

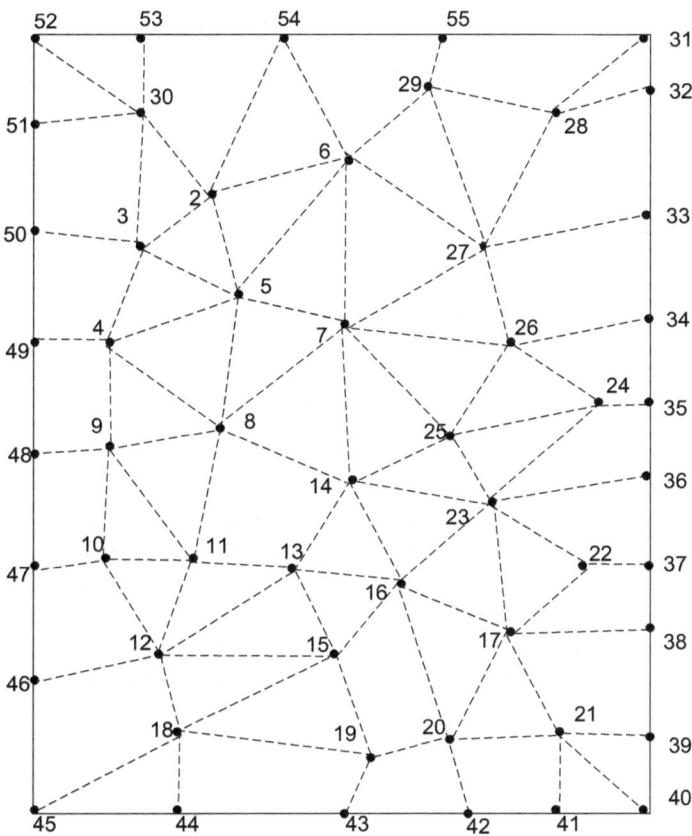

**Fig. 3.6** Graph of a typical area (Makarchi and Tilloston 1991)

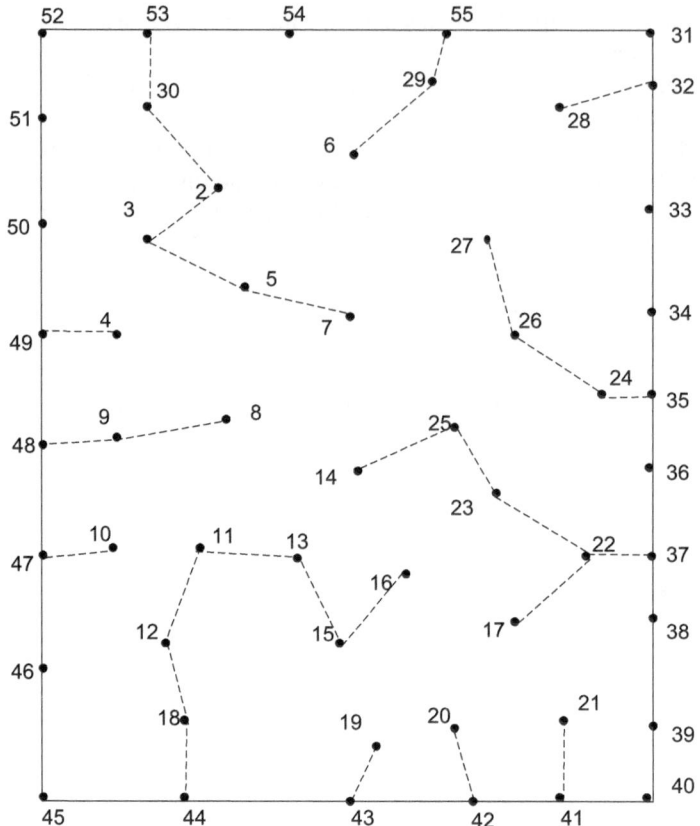

**Fig. 3.7**  MST for the typical area (Makarchi and Tilloston 1991)

The value of Lc will determine whether any network improvement is worthwhile because it influences the trade-off between development and trip costs. A reduction in person-km is a benefit that must be weighed against the expense of increased kilometers. It is only beneficial to switch connections when the benefits outweigh the drawbacks, i.e. when the change in person-kilometers is higher than the product of Lc and the change in kilometers. It can be observed that Lc has a value that determines the break-even point for each conceivable adjustment. Lc can be calculated using Eq. 3.13.

$$L_c = \frac{\text{change in person} - \text{kilometers for the replacement}}{\text{change in kilometers for the replacement}}. \tag{3.13}$$

In the second method, the factor Lc is crucial since it modifies the SST network by accounting for all potential replacements across the district that is being analyzed. For each of these potential replacements, the change in person-km (benefits) and change in kilometers (costs) are listed together with the Lc value, which indicates the replacement's break-even point. The highest available benefit-to-cost ratio (Lc) is then found by reviewing the cost–benefit table. Following the determination of this value, the algorithm executes a second network scan, starting from the node nearest to the destination and proceeding to the farthest village node. Substitutions are made whenever the benefit-to-cost ratio Lc is at least as large as the maximum value that was previously determined. Initially, we anticipate that there will only be a single substitution that matches the highest Lc value discovered. However, altering this connection changes the amount of travels on other links and the route lengths for trips using that link; therefore, multiple replacements may occur and additional replacements may now meet the requirement.

After obtaining a fresh spanning tree network, the procedure iterates, scanning it once more to determine the next replacement(s). With every replacement, the value of Lc associated with that replacement decreases. The process can theoretically come to an end when the value of Lc accurately reflects the trade-off between development and travel expenses. The replacement technique, however, creates the entire sequence of spanning tree networks—starting with the MST and ending with the network that result if trip expenses outweigh construction costs—because the value is unknown.

The computations are connected via recursive equations. Its goal function is to minimize the overall cost, which accounts for construction and transport expenses. Alternative links between linked nodes (root nodes) and disconnected villages are evaluated by the model.

## 3.5   Conclusions

Generally, rural road network models address challenges in flat terrain. The challenges with rural road networks in steep terrains of rural areas differ from those in plain places. In steep locations, settlements and public institutions are sparsely populated. In rural regions, connecting all settlements and public facilities by road is neither practical nor feasible.

Creating a rural road network that connects the majority of communities and public services within a certain radius could be one way to address this difficulty. To do this, the bulk of communities and public spaces must be covered by defined network nodes. We need to establish a basis for identifying nodal points because none of the existing methods outlined above have provided an adequate basis, even in plain terrain. The discovered nodes (obligatory points) can then be linked via rural road networks. This will result in a rural road network with a covering of settlements.

Rural road network prioritization should be determined by straightforward social measures rather than complicated economic figures, with a focus on public facilities, settlement accessibility, and connectivity.

One of the key issues in rural areas is maximizing the coverage of settlements (people) and public services by roads. Another challenge is minimizing transport costs on the roadway network. Problems with the rural road network require more straightforward and workable solutions, especially in rural areas of the hilly region.

# References

1. M. Abdulaal, L.J. LeBlanc, Continuous equilibrium network design models. Transp. Res. B **13**, 19–32 (1978)
2. G. Edmonds, C. Donnges, N. Palarca, *Guidelines on Integrated Rural Accessibility Planning* (ILO/DILG, Manila, 1994)
3. T.L. Magnanti, R.T. Wong, Network design and transportation planning: models and algorithms. Transp. Sci. 18(1), I-55 (1984)
4. Highway Research Board, Highway sufficiency ratings. HRB Bulletin 228 (Washington D.C, 1952)
5. UNCHS (United Nations Centre for Human Settlements), Guidelines for the planning of rural settlements and infrastructure: road networks (Nairobi, Kenya, 1985)
6. C.B. Shrestha, Developing a computer-aided methodology for district road network planning and prioritization in Nepal. Int. J. Transp. Manag. **1**(3), 157–174 (2003)
7. D.L. Oudheusden, L.R. Khan, Planning and development of rural road networks in developing countries. Eur. J. Oper. Res. **32**(3), 353–362 (1987)
8. C.B Shrestha, J.K Routray, Application of settlement interaction based rural road network model in Nawalparasi district of Nepal, in *Technology transfer in road transportation in Africa: Arusha international conference centre*, Tanzania, 2001. Conference proceedings (2002)
9. DoLIDAR, (2010). Approach for the development of agricultural and rural roads: a manual for the preparation of district transport master plan and for the implementation of rural road sub-projects. (Ministry of Local Development, Government of Nepal, Kathmandu)
10. P.K. Sarma, J.K. Routray, D.K. Singh, Spatial analysis of hierarchy of market centres and domestic market potential surface of Central Assam (India). Indian J. Mark. Geogr. **2**(1), 45–64 (1984)
11. W. Isard, *Methods of Regional Analysis: An Introduction of Regional Science* (The Massachusetts Institute of Technology and John Wiley & Sons, Inc., New York, London, 1960)
12. A.K. Mahendru, P.K. Sikdar, S.K. Khanna, Linkage pattern in rural road network planning. J. of Indian Roads Congress **44**(3), 649–675 (1983)
13. J. Howe, Transport for the poor or poor transport? IHE Working paper IP-12, International Institute for Infrastructural, Hydraulic and Environmental Engineering (Delft University, The Netherlands, 1996)
14. G. Edmonds, Wasted time: the price of poor access. RATP No.3 (ILO-Geneva, 1998)
15. K. Dixon-Fyle, Accessibility planning and local development: the application possibilities of the IRAP methodology. RATP No. 2 (ILO-Geneva, 1998)
16. A. Kumar, H.T. Tilloston, "A planning model for rural roads in India", in *Proceedings, Seminar on roads and road transport in rural areas* (Central Road Research Institute, New Delhi, India, 1985)
17. A.K. Makarachi, H.T. Tillotson, Road planning in rural areas of developing countries. Eur. J. Oper. Res. **53**, 279–287 (1991)
18. A.K. Singh, GIS based rural road network planning for developing countries. J. Transp. Eng. (2010). https://doi.org/10.1061/(ASCE)TE.1943-5436.0000212
19. A. Kumar, P. Kumar, User friendly model for planning rural road. Transp. Res. Rec. **1652**, 31–39 (1999)
20. A. Athanasenas, Traffic simulation models for rural road network management. Transp. Res. Part E **33**(3), 233–243 (1997)

21. C.G. Swaminathan, N.B. Lal, A. Kumar, A systems approach to rural road development. J. Indian Roads Congress **42**(4), 885–904 (1982)
22. R.D. Mackinnon, M.J. Hodgson, *The Highway System of South Western Ontario And Quebec: Some Simple Network Generation Models* (University of Toronto, Urban environment study, 1969)
23. C.B. Shrestha, Manual for the preparation of a District Transportation Master Plan, Pilot Labour Based District Road Rehabilitation and Maintenance Project (Butwal, Nepal, 1997)
24. G.B. Dantzig, S.F. Maier, Formulating and solving the network design problem by decomposition. Transp. Res. B **13**, 5–17 (1978)

# Chapter 4
# Rural Road Construction in Local Conditions

## 4.1 Introduction

This study focuses on rural road development, particularly in steep rural areas of developing countries. This chapter provides a brief overview of rural road development in mountainous areas of developing countries.

Road construction is often linked to the geological and geotechnical characteristics of their geographic location. The degree of these characteristics is especially important when discussing slope stability and other geotechnical issues that are common in hilly areas. This will lead to a description of the geography, geology, and climate history of numerous areas. Landforms and erosion processes are used to explain the geomorphology of a mountain model [1] in relation to road building in mountainous areas of developing countries.

Since they are directly related to many facets of geology, slope stability, and drainage, earthquakes, volcanic activity, erosion, and rainfall are significant natural elements that have an impact on road construction on steep mountain slopes. The more severe these conditions are, the more challenging road construction issues become.

In mountainous regions, the most significant geotechnical engineering issues are associated with the properties of the soil and rocks close to the surface, drainage, and slope stability. The sections that follow provide a brief explanation of some of these characteristics: tectonic movement, topography, climates, geography, and geology.

When developing rural roads, geographical factors such as the distribution of settlements must be taken into account in addition to technical considerations. This is also briefly discussed.

J. Shrestha, *Rural Road Development in Developing Countries*,
SpringerBriefs in Applied Sciences and Technology,
https://doi.org/10.1007/978-981-96-2012-8_4

### 4.1.1  Tectonic Environment

Tectonic plate boundaries and mountain belt evolution have a close relationship. The history of earthquakes in these belts suggests that the mountainous regions are still tectonically active. An additional element that is crucial for road construction is that seismic activity frequently acts as a catalyst for catastrophic erosion and widespread slope instability, which in turn causes a rapid increase in the sediment loads in mountain streams. The significant nature of the tectonic disturbances for road construction is indicated by the total scale of the mountain relief and the intricacy of the rock structures generated. In regions with high relief, slope erosion rates are often at their highest.

The largest mountain range of the world, the Himalayas spans around 2400 km in total. The Nepal Himalayas cover an area of about 800 km and are located in the middle of the Himalayan Arc, between the Kumaon Himalaya to the west and the Sikkim-Bhutan Himalaya to the east. The sub-tectonic units of the Nepal Himalaya are the Tibetan-Tethys Himalaya, the Lesser Himalaya, the Higher Himalaya, the Indo-Gangetic Plain, and the Sub-Himalaya (Siwalik Group), which stretch from south to north. The Himalayan region is separated by nearly east–west thrust systems that divide the various major geological units. From north to south, these thrusts are the Main Central Thrust, Main Boundary Thrust, Main Frontal Thrust, South Tibetan Detachment System, and Indus-Tsangpo Suture (Fig. 4.1).

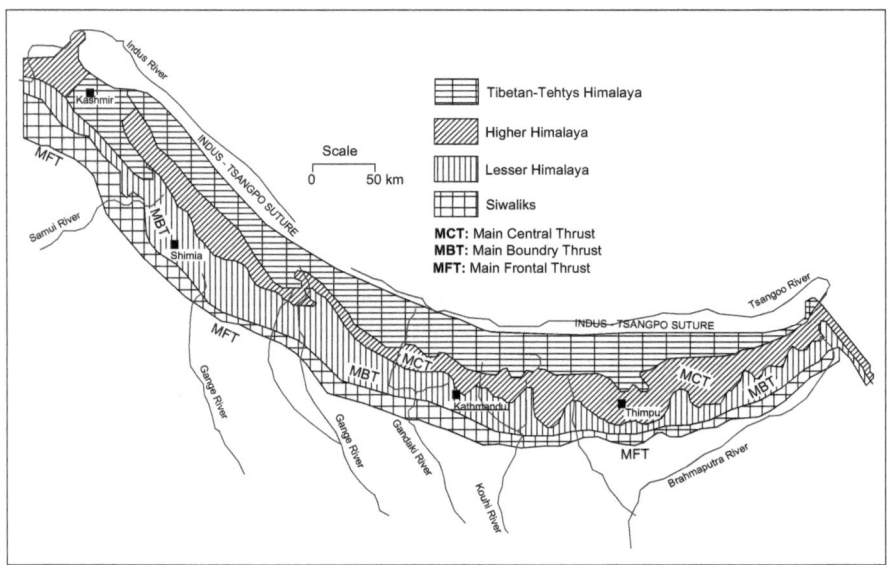

**Fig. 4.1**  Longitudinal geological subdivision of Nepal Himalaya [2]

## *4.1.2  Geology*

The near-surface geology of the mountain belts is incredibly varied, making it impossible to draw any firm conclusions regarding the distribution of common rock types and structures. Tectonic upheaval from clashing plates generated gravity slides, which pushed large amounts of rock down slope and toward the boundary of the mountain zone. The original sedimentary layers were uplifted, folded, and contracted laterally as a result of this process, forming the mountain ranges.

The mountain belts often contain two large suites of rock types as a result of these mechanisms. Deeply seated igneous and volcanic rocks coexist with the initial sedimentary deposits, which are primarily thick shales, siltstones, and sandstones, in the central core. The bulk of the initial sedimentary rocks have undergone varying degrees of metamorphism and are now slates, schists, and gneisses with basic and intermediate lavas and pyroclastic volcanic rocks. Sandstones and limestones, on the other hand, are examples of the mildly metamorphosed sedimentary rocks that define surrounding regions. These are frequently covered in or buried beneath other igneous and metamorphic rocks that have been carried from the central zone by nappe structures and gravity slides. As a result, lithology and structure across a major fold mountain region vary greatly, and a variety of transported and in situ residual soils may cover the outcrop pattern of the different rock types.

Important geological components for engineering include the different kinds of rocks, the weathering and soil formation structures and patterns, slope erosion, and instability features. On a regional scale, the type of rock may be the most significant factor influencing slope stability. [3] investigated natural instability on five different types of rock in Nepal. Gneiss and quartzite slopes were shown to be significantly more stable than schist, shale, and phyllite slopes. In reality, there is rarely a homogeneous rock structure, topography, or weathering depth along a significant portion of the mountain railroad. It might not be practical to choose an alignment of a road solely based on the types of rocks present.

Sedimentary soft rocks, including mudstones, shale, sandstones, siltstones, and conglomerates, make up the Siwalik (Churia) Range. These rocks are brittle and readily broken up. Thick beds of loose, brittle conglomerates can be found in the Upper Siwalik. Similar issues with shifting mudstone and sandstone strata exist in Lower Siwalik and Middle Siwalik. Mudstone in these alternate bands has a tendency to swell and flow when wet, causing sandstone beds to overhang. Blocks made of such overhang jointed sandstone strata break quickly. The Churia Range typically receives 2000–2500 mm of precipitation annually. Because of this, the climate and geology of the Churia Range make landslide processes extremely likely. In Siwaliks, rock collapses, shallow slides, and debris flows are basically the typical outcomes of weathering.

The Lesser Himalayan Zone includes the Mahabharat Range. It is the primary monsoon cloud barrier and has a significant impact on rainfall distribution pattern in Nepal. Compared to Midland, the southern face of the Mahabharata Range receives more rainfall almost everywhere in Nepal. In the Mahabharat Range region, there is a higher yearly rainfall as well as a higher frequency of intense rainfall. As a result, these regions frequently experience shallow landslides, debris flows, and flooding.

In addition to rainfall, the Mahabharat Range's extremely steep slopes and geological characteristics also play a significant role in soil slides and debris flows. It is observed that the slopes of the Mahabharat Range are more stable in the region composed of rocks like granite, dolomite marble, and limestone. However, the topography is particularly vulnerable to landslides in the area with rocks including phyllites, slates, and quartzites and phyllites intercalated.

One of the main activities in the lower Himalayas that disturb the landslide-prone mountain slopes is constructing rural roads. Thus, the risk can be reduced by carefully selecting the alignment and design of the road.

### 4.1.3  Topography

The magnitude of relative relief and its relationship to the dissection system's density are the most significant features from a geological perspective, since they largely determine the steepness and stability of the slope. In addition to the impact of climate and rock type, topography plays a significant role in the formation of mountains, as rates of uplift and denudation change over time. [4] was the first to offer the plan for developing mountain landscapes. His approach operates on two assumptions: first, that rivers and streams have two phases of activity, and second, that uplift significantly exceeds denudation until the mountain mass is tectonically stable.

Rapid incision or down-cutting occurs during the early phase as rivers actively clear their channels. When the main rivers have developed flood basins and are well acclimated to baseline levels of erosion, down-cutting is curtailed in the second phase. The river system creates a disjointed alpine landscape with steep valley-side slopes, a drainage network that is incising, and many naked rock outcrops. Typically, these slopes are made up of a sequence of nearly straight sections that, as glacial activity increases, become steeper and retain their high angle to the main river edge of the valley.

After a river creates a floodplain, average slope angles steadily decrease as the valley-side slopes get depleted, and slope growth is mainly independent of river activity. In specifics, differing rock types, rates of erosion greatly influence how a mountain slope looks. Numerous geotechnical issues in road construction can be attributed to the powerful erosion processes of surface water erosion and landslides in these high relief places.

### 4.1.4  Climate

Because altitude alters regional climate patterns and creates broadly vertical zones of climatic type, mountain climates typically differ significantly over short distances.

Within this larger framework, localized differences in slope concentration, relative relief, and slope angles give rise to unique microclimates. Climate controls the relative rate of denudation, which affects how mountain slopes evolve. Changes in the average yearly temperature and precipitation have an impact on the relative importance of various weathering and erosion processes.

The change of rainfall with elevation is more complex than the drop in temperature, which is roughly linear at the regional scale. Rainfall generally rises with elevation to a point, usually between 1000 and 2000 m, however, then falls at higher elevations. On the other hand, the effects of the local climate and geography are also significant. Most monsoon rains fall on the southern windward side of the Himalayan foothill ranges; the amount of rain reduces sharply on the northern side of each successive range as altitude rises [5, 6]. Anywhere there are mountain ranges acting as a barrier to rain-bearing winds, these rain shadow effects are typical. For a road routing along the rain shadow slopes, it would be advantageous.

Rainfall is typically the most significant climatic indicator when it comes to geotechnical construction issues. The annual total rainfall isn't always the most crucial factor, though. According to Radbruck-Hall and Varnes [7], when seasonal rainfall exceeds 250 mm, there is a significant increase in the frequency of slope failure in temperate climates.

## 4.1.5 Geography

The topography of Nepal is split into three regions that span its whole length: the Terai, or southern plains, the Mid-Hills, and the Mountains (Fig. 4.2). The geography of Nepal includes some of the world's most varied climatic ranges and natural habitats. A distance of 170 km separates the highest point on Earth, Mount Everest, at 8848 m, from the Gangetic plains in Terai, which are 70 m high above sea level [8]. With rich lowlands, the Terai region, which makes up 23% of the overall area, is roughly 30 km wide and ranges in elevation from 70 to 280 m. From the end of Terai to the Himalayas, the topographical slope thereafter gradually increases. These are the steepest slopes in the globe, with multiple streams, tributaries, and wild rivers dividing numerous villages.

Numerous rivers intersect with trails that wind through the untamed landscape. Alongside these topographic extremes are temperature extremes, ranging from − 35 °C in the Himalayas to 48 °C in the tropical Terai. The summertime melting of snow combined with extremely high monsoon precipitation from higher alpine altitudes lead to the swelling of rivers beyond control. 87% of the country's area is made up of hills (52% of the total area) and the Himalayas (25% of the total area).

In the steep regions, settlements are dispersed throughout the hillside. Settlements on the ridge slopes are typically denser. Instability features like landslides and erosions around the streamlines, which are typically covered with forests. Since hill slopes are unstable and have steep slopes, there is less population in valleys where streams flow. The majority of valley slopes remain uncultivated because they are

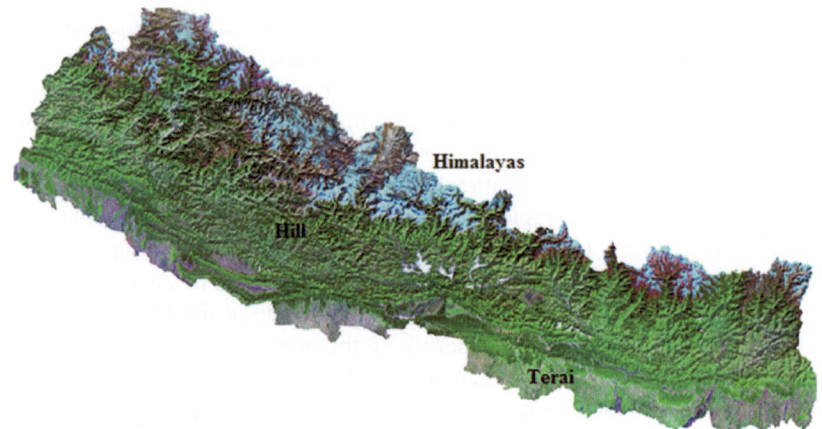

**Fig. 4.2** Physiographic features of Nepal [9]

difficult to farm. In the hilly regions, the stable terrain is either cultivated or occupied for settlement. Therefore, stable places are those where people have lived and farmed for a long period. Settlements are therefore concentrated on the higher hill slopes since instability issues are less likely to arise there. Along with the settlements, the public infrastructures are located in the same area near the ridgelines on the hill slopes.

Most settlements and public spaces are connected by road alignments that cross ridges of hills. To connect or offer access to the towns and public services, roads must be laid out as much as feasible along the ridgeline. However, locating ideal alignments might not be easy. Road alignments typically cross near ridgelines. Locating rural road alignment along hill slope ridgelines is advantageous since the population is concentrated along these areas and the slope is steady close to these ridgelines.

## 4.2   The Five-Zone Mountain Model

Numerous landforms and associated geological and geotechnical issues result from the topography, geology, and climate variation found in mountainous regions. Five primary zones comprise the basic concept of fold mountain topography that [1] presented. Below is a full description of the model's units, which contain typical land structures, materials, and weathering processes related to each zone, with a focus on drainage, river crossings, earthworks, and slope stability. The model is presented in Fig. 4.3.

Zone 1: The High Himalayan region's typical glacial and periglacial morphology. The landscape of the high peaks is made of rock and ice.

Zone 2: The High Himalayas and the highest elevations of the Middle Himalayas are known for their free rock faces and the slopes of debris that accompany them. Zone

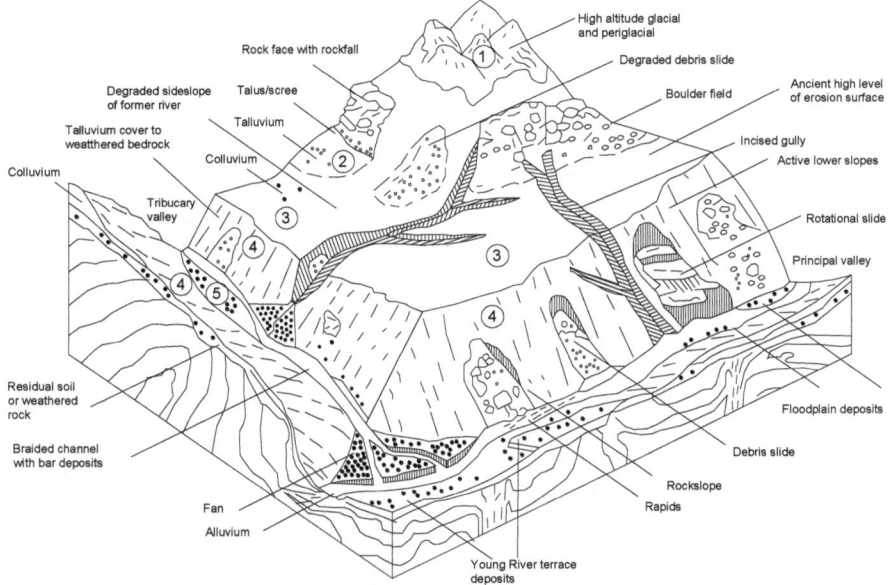

**Fig. 4.3**  Model for young fold mountains [1]

2 comprises coarse debris-coated slopes and fragments of immature rock. Common landforms include exposed rock peaks, cliffs, ridges, screes, boulder cones, and steep slopes covered with coarse weathering debris. Zone 2 scree goes all the way to the bottom of the valley. Most rocky outcrops in mountainous regions are made of strong jointed rock and susceptible to several instability mechanisms. The most common are rockfalls, wedge, and toppling failures, weather-related, small-scale events that happen on rock faces.

Larger material quantities are often involved in slab collapses, rock avalanches, and rock slides, which are frequently connected to significant joints and shear planes.

Zone 3: Ancient valley floors and degraded middle slopes; these features are typical of the Mahabharat Lekh and the lower elevations of the Middle Himalayas. Zone 3 topography differs from Zones 2 and 4 in that it has relict landforms that show the history of denudation of the area, a thicker soil cover, and generally lower average slope angles. This zone's key characteristic is that, compared to Zones 2 and 4, there are far fewer issues with slope stability and less stringent restrictions on the alignment and cross-section of the road. There might be sources for construction materials in the zone.

These are typically transported soils and in situ residual soils, notwithstanding the possibility of encountering a broad variety of soil and rock materials. The word "cluvivium," which is used loosely to describe the features of transported slope debris, usually refers to larger-sized particles of gravel and other rocks that are bonded together in a clayey matrix at different phases of weathering. It generally exhibits a wide range of cohesiveness and frictional strength, which reflects the variable character of the material throughout more advanced weathering stages. Fine colluvial soils can have extremely low residual strengths. As a result, if the slope has previously been unstable and has relic shear surfaces, cuttings, and foundations need to be carefully evaluated.

There are two main types of residual soils: fully weathered soils that are devoid of structure, and soils that retain the remnant rock structure, called saprolites, which are found in weathering grades 4 and 5. In saprolitic soils, discontinuities persist as areas of relative weakness. It is possible for quick chemical weathering along with preferential groundwater circulation along primary joint planes to generate clay and other weak infills or coatings, which have very low strength.

These infills have the potential to provide barriers to groundwater migration and significantly reduce permeability along joint planes over time. Perched water tables and issues with soil instability above can result from this. Zone 3, which has a more complete succession of the different weathering grades, has a much thicker weathered rock profile beneath the soils than the more active erosive slopes in Zones 2 and 4.

Zone 3 slopes are rather infrequent in terms of natural slides and mass waste. Excavation-related failures mostly affect the deep residual and colluvial materials on the side slopes of ancient valleys. These are frequently linked to the reactivation of previous slides that were started in Zone 4 terrain or the emergence of locally elevated groundwater levels. This geographic expression of the old slides is frequently subdued and masked by vegetation. In several instances, their identification necessitates meticulous groundwork exploration throughout the reconnaissance and/or site investigation phases. On Zone 3 slopes, surface water erosion and slow soil creep are the two main natural erosion processes. For bare slopes in both colluvial and residual soils, especially those with fine sand or silt grading, rilling and gully erosion are key challenges. When it rains heavily, surface and subsurface water can cause considerable raveling in the thin colluvium layers at the top of the cut slopes. The colluviums are frequently more porous than the underlying residual soil or weathered rock.

Zone 4: Active lower slopes, comprising much of the Mahabharat Lekh and portions of the more restricted slopes in the Middle Himalayas near the major rivers. The active valley-side slopes of Zone 4 make up a large portion of the terrain. The terrain is characterized by side slopes of gullies and valleys that slope sharply and are covered in a mantle of transported soil on top of weathered bedrock. These slopes have high rates of erosion; therefore, construction in this area typically presents the most severe challenges.

Landforms like Zone 2—geodes and steep angled rock slopes—occur in areas subject to significant river undercutting. A general state of possible instability prevails if the average zone slope angle is more than around 30°, and many of the steeper soil-covered slopes will have factors of safety that are marginally above unity.

On Zone 4 slopes, stability and drainage issues have a significant influence on the selection of alignment and cross-section. Shallow landslides and gully erosion occur often across a wide range of rock types and climate conditions. There are three primary forms of shallow landslides: debris slides, rockslides, and mudslides. Mainly the higher layers of transported and residual soils are where debris slides occur.

On steep slopes covered in soil, natural instability is mostly caused by intense rains. However, the effects of all-natural factors are typically less significant during construction than the over-steepening caused on by excavation. In a rockslide, more material is usually involved, and the slide surface is often located beneath the soil cover in the area of weathered rock or along discontinuities that are orientated unfavorably at the surface of the low-weathered rock. Mudslides are mainly restricted to fine-grained rocks and soils that have seen intense penetrative weathering. The presence of perennial near-surface groundwater is often associated with them.

Classical rotational slides and flow failures occur less commonly. At deeper depths, shear surfaces set rotational slides apart. These slides usually occur in heavily worn rock slopes where the lack of strength in the rock mass from weathering outweighs the influence of rapidly moving streams full of debris and moisture. They can also occur when erosion has undercut slopes in deeply deposited soils. Typically, flows arise in weak rock and natural soils. They can also originate when surface runoff or heavy rain concentrates on massive spoil tips created during earthwork excavations.

Zone 4 slope construction is impacted by surface water erosion mostly due to gullies and rills. A mixed bed load and a brief peak flow associated with significant rainstorms are prominent characteristics of gully erosion. Sediment concentrations are degraded and unevenly redistributed throughout the gully floor at each major flood.

Actively eroding mountain gullies are characterized by irregular steep longitudinal profiles ranging from 10–45 , and a usually V-shaped cross-section with a steep side slope that descends to the channel floor. Since gully crossings are significant expense items, the density of the drainage network in Zone 4 may be several times higher than in Zone 3 for a given amount of drainage channel. This is an important consideration when choosing an alignment. The drainage pattern has a significant impact on alignment selection as well.

In some fine-grained rocks and soils, rilling occurs on extreme naked slopes with no vegetation. Unlike gully erosion, it is more frequently linked to human intervention through the removal of natural vegetative cover through spoil tipping. In the same way, subsurface erosion on naturally occurring slopes seldom poses a direct threat to the development of mountain roads because most engineering problems occur after construction.

Zone 5: Consists of valley floors that are typical of the Middle and Lower Himalayas and, to a lesser extent, of the former. The tributary stream and gully crossings at the point where they pour into the main river valley are the main points of interest. Road alignments that parallel the edges of large rivers are frequently chosen in mountainous regions where building would be difficult and expensive due to high flood velocities and sediment loads, mainly in Zone 5 terrain. These tributary stream exits frequently experience sudden changes in bed gradient and channel width, and sediment deposition prefers to occur here. The distinctive features of depositional landforms are alluvial fans.

Fans are particularly common in semi-desert and monsoonal regions, where stream flow is ephemeral, and where the ratio of depositional area to mountain catchment area is minimal. Typically, fans are composed of gravel-clay-sized debris. Single fans typically have a maximum slope angle of 10° and a conical shape.

Cobbles and boulders, which are coarser sediments deposited by debris flows, are also linked to fans at the base of unstable mountain catchments. The bed level of a stream can change by several meters in hours or sometimes even minutes due to sediment deposition that occurs during a quick erosion episode on Zone 4 slopes. The main river channel will be diverted by a quickly expanding fan, which will cause erosion on the other bank.

In mountainous terrain, constructing rural roads is the main activity. The planning and building of rural roads in Nepal's hilly regions can benefit greatly from the application of the five-zone mountain model since the construction of roads is closely linked to the regional geology along their alignment. It offers guidelines for choosing the alignment of rural roads and designing their cross-sections.

## 4.3  Alignment Selection and Choice of Cross-Section

Any road corridor should be located with a minimum grade and as few river crossings and materials as possible. Zone 3's mature slopes, firm terraces, and plateaux provide the ideal route. The main guideline of corridor location applies when it is necessary to cover all zones: make height in Zone 4, make distance in Zone 3, and avoid running prolonged lines across slopes in Zone 4. Finding safe zones and connecting them with stable corridors and river crossings is the alignment engineer's expertise. Rural roads typically travel through Zones 3 and 4.

Horizontal curves may have a minimum radius of less than 20 m, with the largest permitted gradient usually being 1 in 10. The prevailing gradient is usually of the order of 1 in 15. Applying these requirements typically results in a mountain road that is two to four times longer overall than the planned length between the end locations. Slopes for cutting may be as high as 35 m, and they may even be higher in Zone 5 George areas and steep rock slopes on Zone 2. According to [1], retaining structures may be needed up to a height of approximately 15 m and may span 20% of the whole length of the road.

This rule is equally applicable to rural road alignment selection. Nevertheless, the geometric standard of road is substantially less than the previously mentioned figure. The road geometry for country roads has been further loosened by the rural road standard [10], which sets the minimum radius of curvature at 10 m, maximum gradient at 12%, and maximum average gradient in hills at 8%.

In mountainous terrain, there are two conventional methods for choosing a cross-section: (i) the oldest method, which is still in use, is to use a full-cut profile with side tipping of spoil throughout; (ii) the second method is to apply the cut-and-fill balance principle, which is taken from conventional road construction. These two methods are not suitable for widespread use. The method chosen needs to account for the changing slope forms and properties of the rock and soil. On country roads, however, the sensitivity of the cut mass is crucial. The ecology is greatly impacted by the chopped mass. The zonal model can serve as the foundation for a logical cross-section design strategy.

Two broad guidelines apply if stability issues are met: stay away from rock cuts and structures, both of which are expensive, unless there is an obvious better option. On hill slopes higher than roughly 30°, retaining structures are necessary because many common fill materials will become unstable at this angle, keying in the base of an embankment becomes expensive, and overburdened soils are more likely to slide shallowly. However, it would be feasible to employ well-founded and well-placed rockfill on slopes up to about 45°. There will inevitably be rock cuts on slopes higher than 45°. When there is a significant excess or deficit of material on potentially unstable slopes, the only appropriate cross-section can lead to earthworks issues.

Burrow and spoil disposal sites should be placed on such slopes with the same consideration as the alignment itself.

The most stable and advantageous alignment from an economic standpoint is Zone 3. Alignment issues mostly arise near the edges of the zones. The choice of where to enter and exit Zone 3 is the most crucial. Generally speaking, managing instability from below the line is more challenging than handling instability from above. Rockfall poses a threat to the alignment at the higher border. Major scarp face retreat—which is particularly difficult to prevent—and localized gully down-cutting—which is easier to control—may result from instability at the lower boundary Zone 4.

The ridgehead route, sometimes referred to as the watershed or spine route, is a unique example of a Zone 3 alignment that is completely surrounded by Zone 4. It has been extensively adopted.

By using a cut-and-fill balance, the alignment within Zone 3 will aim to minimize costs. In most cases, cuts and fills will be shallow; however, in cases where the rock head is shallow as well, the vertical alignment needs to be carefully evaluated in order to prevent rock excavation whenever feasible.

Finding stable ascending paths to connect Zone 5 river crossings with Zone 3 slopes, when appropriate, is the alignment engineer's task in Zone 4. A common and contentious practice when utilizing small climb corridors is the vertical stacking of hairpin loops, sometimes known as switchbacks or zigzags.

A bank of hairpin bends is especially prone to increasing instability, which can arise from even a little failure in fill or cut and occur either upslope or downslope. Stabilization, drainage, erosion management, and retaining structure construction come at a high expense. The usage of hairpin bends, however, shortens the length of the road overall and can save money as compared to locally costly construction at hairpin locations. To lessen the impact of cumulative failures, the hairpins can be divided into smaller groups and offset from the groups directly above and below if the ascent corridor is sufficiently wide.

Hairpin loops can be arranged one above the other in road alignments so that water diverted from the uphill road does not fall on the downhill road segment. In the event of an uphill road failure, this might prevent the road segments from failing gradually.

For steep Zone 4 slopes, there are two design approaches for cross-sections. First, reduce the impact of shallow cuts and fills in the earthworks; second, utilize sound rock to the fullest extent possible at depth. Walls and revetments are widely utilized as a compromise between these two opposing objectives. Because the cutting excavation frequently exposes rock for a stable formation, retained cut is preferred over-retained fill. On the other hand, the amount of earthwork needed on country roads should be kept to a minimum because deep cutting would need a lot of excavation that would be challenging to manage. For country roads, the first principle is appropriate.

Due to its significantly lower slope angles, Zone 5 is typically chosen over Zone 4 for alignments; yet, Zone 5's high flood levels could very well be the deciding decision. These Zone 5 roads are longer overall and have less geotechnical risk, but they require substantial and costly development.

The most costly components of Zone 5 alignments are bridge crossings and erosion control measures. In order to minimize tributary crossings and prevent river undercutting, active debris fans, slides, and cliffs, it might be essential to cross the mainstream. However, alignment in Zone 5 is not advised for rural roads due to the unpaved surface of most country roads and the need for numerous cross-drainage devices. Rural road construction in this area comes at a high cost.

Generally speaking, a complete fill cross-section will be used in Zone 5 to prevent cutting into the slopes in Zone 4, with the exception of situations in which the encroaching embankment poses more serious issues than in Zone 4. To reduce embankment construction and make maintenance access to the embankment face easier, the best altitude for the pathway is slightly higher than the high water mark.

## 4.4   Impacts of Roads' Construction on Slopes

The slope of a mountain can be significantly impacted by the construction of roads. There has been significant soil loss in Nepal as a result of rapid erosion, gullying, and landslides brought on by the vast clearing of mountain slopes, the negligent disposal of the cleared materials downstream, unregulated rock blasting, quarrying, and mining operations, and inadequate water management. Development of a road could result in the excavation of between 100,000 and 200,000 tons of material every kilometer. For unprotected cut slopes, an extra 100 tons of slide materials can be added every kilometer each year. Because mountain roads were constructed haphazardly 8000 metric tons of soil per hectare is lost annually.

[11] reported that insufficient maintenance of poorly constructed mountain roads results in an annual loss of 1000 metric tons of soil per hectare.

Typically, the most significant climatic indicator is rainfall. The annual total isn't always the most crucial factor. When seasonal rainfall exceeds 250 mm in temperate locations, slope failure occurs more frequently [5]. Due to this, landslides on mountain slopes occur frequently during the monsoon season. Based on estimates, landslides along mountain roads occur 400–700 cubic meters per km per year, while during the construction of mountain roads in Nepal, landslides occur

3000–9000 cubic meters per km [11]. 500 cum/km/year of debris can be produced on average by the cut slope failure following construction. In Nepal and India, single storms with repeated intervals of 10–20 years can produce up to 2000/cum/km/year [12].

This demonstrates the tremendous environmental impact of road construction in hilly terrain of Nepal. To mitigate these effects, the cut volumes in road cross-sections must be lowered. This is only achievable by reducing the cut height and breadth, which will require less excavation.

This leads to instability in the hill slope as the cutting height increases along with the breadth of cuts. The stability of valley slopes and the loss of agricultural and forest areas are two major environmental effects of the cut and throw of excavated materials. Consequently, the cut slope should be made much more stable and safe by designing the road cross-section with the cut slope height as low as possible. It is recommended to utilize as much of the excavated materials as feasible for construction purposes and to properly dispose of the leftover materials to prevent erosion and landslides caused by rolling downhill.

These illustrate the requirement for cut-and-fill sectional areas to improve cut mass management by making the most use of excavated materials (Fig. 4.4). Stone blocks that have been excavated can be gathered and used to build retaining walls.

In hilly areas, a study found that slopes between 21 and 40 make up 55.33% of the entire road length and contribute 53% of the cut volume per kilometer of road construction [13].

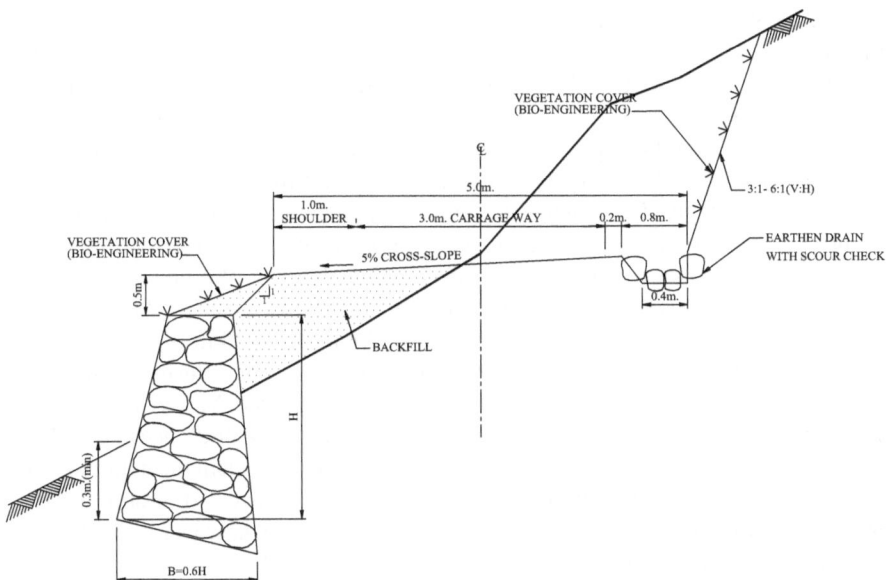

**Fig. 4.4** Typical cross-section in cut and fill [14]

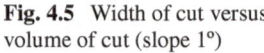

**Fig. 4.5** Width of cut versus
volume of cut (slope 1°)

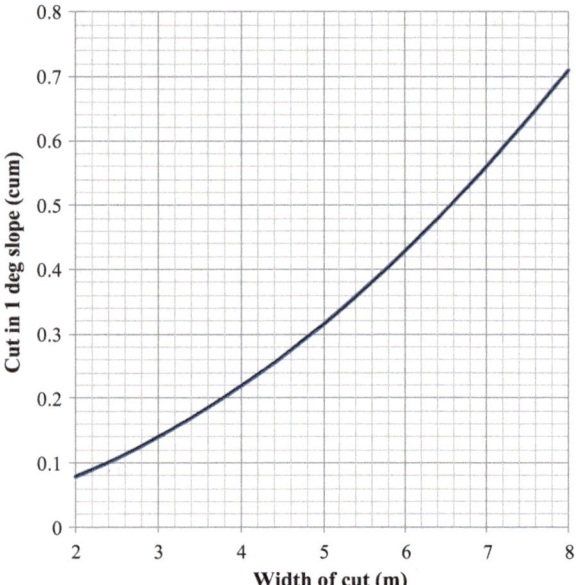

The cross-sectional excavation volume is affected by the slope of the mountains
and the breadth of the cut (Fig. 4.5). The volume and the breadth of a cut have a
quadratic relationship when the slope is included. The quadric relationship between
the volume of cut for a given width of cut and the hill's slope grows correspondingly
(Fig. 4.6).

This study shows that in hilly places, the amount of cut volume in rural roads
is particularly sensitive to the cut breadth and slope of the terrain. The surrounding
ecology is significantly impacted by the increased amount of cut. Therefore, it is
important to work on lowering its volume. The cut-and-fill cross-sections (Figs. 4.4
and 4.8) should be used as much as possible for this exercise; cross-section Fig. 4.7
should be avoided. As a result, slope stability is minimized by reducing the
disturbance to mountain slopes and the cut width and height of road cross-sections.

In the hilly regions of Nepal, rural roads are among the most sought-after development infrastructures. According to [10], Nepal has 32,580 km of rural roads, of
which only 10,000 km are used. Every year, from June to August, during the monsoon
season, the majority of the hilly region's rural routes are closed. Moreover, landslides
and poor road surface conditions generate rural road network bottlenecks during the
monsoon season, cutting off the majority of highland regions from the national transportation network. A substantial sum of money has been invested in building rural
roads, especially in the highlands. Planning and engineering, however, were the areas
with the greatest shortcomings.

Thousands of kilometers of road were built with little concern for proper planning, engineering, or environmental implications. As a result, only a few roads are
operational. These roadways have exacerbated slope instability issues, resulting

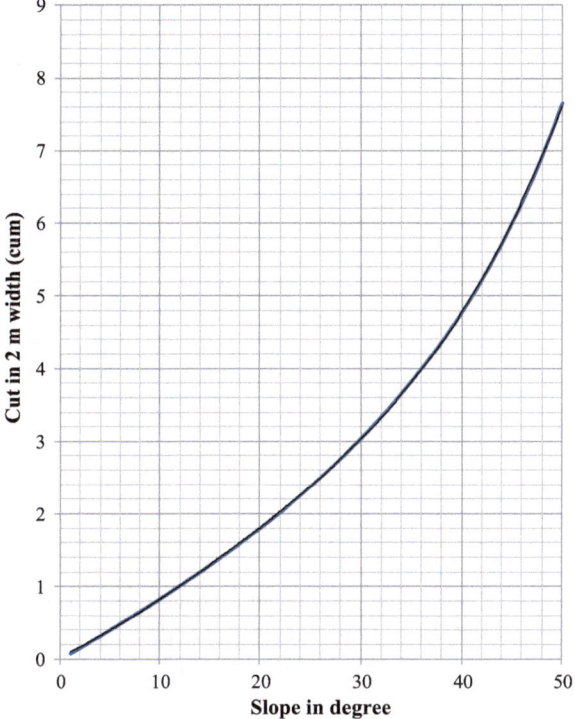

**Fig. 4.6**   Slope versus volume of cut (2 m width)

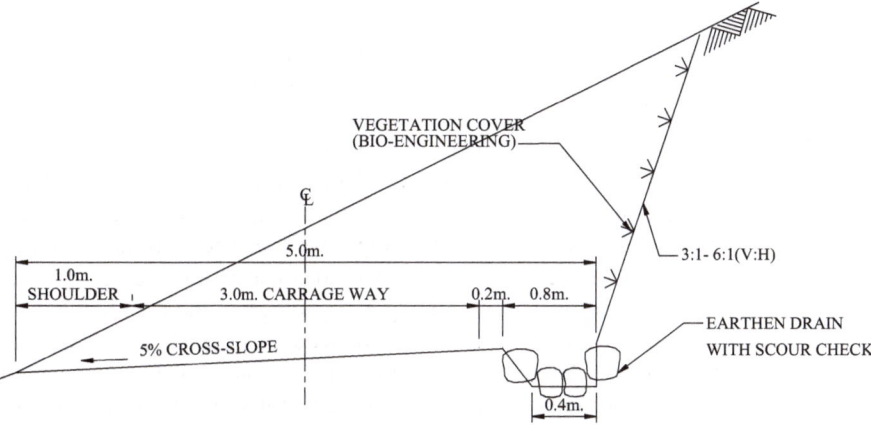

**Fig. 4.7**   Typical cross-section in full cut [14]

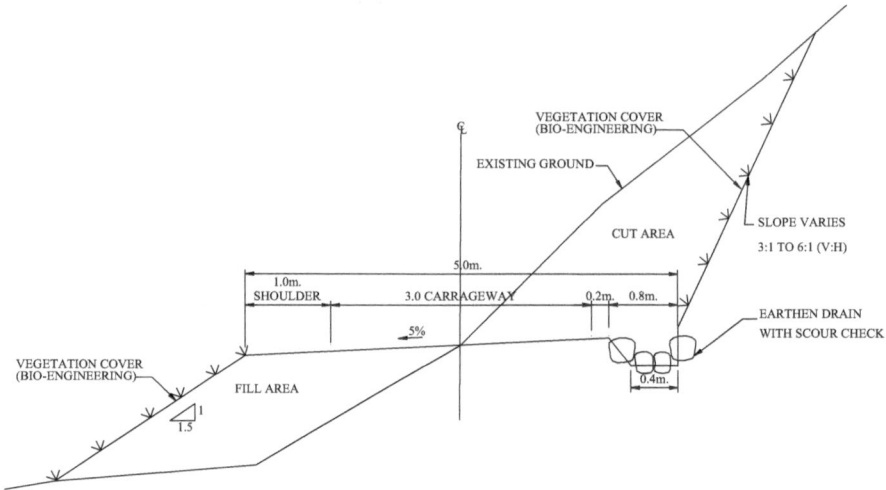

**Fig. 4.8** Typical cross-section in mild slopes [14]

in several landslides. Even under engineering surveillance, roads are constructed without considering instabilities both during and following construction. The poor geological materials and the region's topography, which has many rivers, streams, and water drains as well as steep hills, are to blame for this.

It is vital to be knowledgeable of the local geoenvironment, particularly in the hilly regions of Nepal where infrastructures are to be created. The hilly regions of Nepal are situated in the Himalayas, one of the world's most active fold mountain systems with high levels of erosion [11]. The problems caused by unfavorable geology, topography, and land use are worsened in humid subtropical and warm-temperature regions where high rainfall and fast rock weathering lead to land erosion and sliding. Therefore, the construction needs to consider the unique conditions of the area.

In Nepal, knowledge of the geomechanical behavior of Himalayan geological materials is relatively limited. The expansion of infrastructure has therefore led to expensive building and maintenance expenses, frequent damage from rugged terrain, intense monsoon rains, and an extensive building schedule. The surrounding topography and geology are significantly impacted by road construction. Numerous roads built without consideration for the topography and geology of the area had a major after-construction effect on the nearby ecology. Road development has caused significant soil erosion, which has led to slope instabilities in the natural slope integrity on the fragile geological conditions in road corridors. This illustrates how the construction of new roads can significantly affect drainage, erosion, slope stability, and the flow of silt into drainage networks. Moreover, its knowledge is important before harming the local ecology.

# 4.5  Conclusions

In the hilly landscape of Nepal, the five-zone mountain model works well for identifying and evaluating rural road alignments. The zonal model can be used to analyze geological and geomorphological processes, as well as the repercussions of road construction. In Nepal's hill, most rural roads travel via corridors designated as mountain Zone 3 and Zone 4. In hilly terrain, mountain Zone 3 is considered the most durable area to locate and build rural roads. The hill slopes of most rural roads are between 20 and 40 , with valley-side retaining structures and cut-and-fill cross-sections.

Additionally, in hilly terrain of Nepal, the post-construction effects of road construction on the environment can be very significant because of the excavated materials and the construction of mountain slopes vulnerable to landslides. Cut-and-fill road cross-sections must be provided to minimize the cut width during engineering design and construction since the cut width in a hill slope is especially sensitive to the volume of quantity. To minimize the amount of excavation when creating a road, careful consideration should be given to selecting cross-sections that are appropriate for the slope of the terrain.

The majority of communities and public amenities are found close to ridgeline slopes (Zone 3). The demand for road alignments passing through this zone and stable slopes are correlated with this occurrence. This fact provides a significant backdrop, both technical and social, for developing a rural road network design model for the hilly regions of Nepal.

# References

1. P.G. Fookes, M. Sweeney, C.N.D. Manby, R.P. Martin, Geological and geotechnical engineering aspects of low-cost roads in mountainous terrain. Eng. Geol. **21**, 1–152 (1985)
2. A. Gansser, *Geology of the Himalayas* (Interscience, Wiley, New York, 1964), pp.1–289
3. D. Brusden, D.K.C. Jones, R.P. Martin, J.C. Doorncamp, The geomorphological character of the Low Himalaya of Eastern Nepal. Z. Geomorphol. Suppl. **27**, 25–72 (1981)
4. W.M. Davis, The Geographical Cycle. The Geographical Journal, 14(5), 481-504 (1899)
5. J.L. Nayava, Heavy monsoon rainfall in Nepal. Weather **29**(12), 443–450 (1974)
6. O.N. Dhar, B.K. Bhattacharya, Variation of rainfall with elevation in the Himalayas—a pilot study. Indian J. Power River Val. Dev. **26**(6), 179–185 (1976)
7. D.H. Radbruch-Hall, D.T. Varnes, Landslides—cause and effects. Bull. Int. Assoc. Eng. Geol. **14**, 205–216 (1976)
8. Central Bureau of Statistics (CBS), (2011). Nepal in Figures (National Planning Commission secretariat, Government of Nepal, 1899)
9. Rural Access Improvement & Decentralization Project (RAIDP), Preparation of GIS based rural roads network and accessibility mapping at district and national levels based on the District Transport Master Plans (DTMPs). DoLIDAR (Ministry of Local Development, Government of Nepal, Kathmandu, 2009)
10. DoLIDAR, Approach for the development of agricultural and rural roads: a manual for the preparation of district transport master plan and for the implementation of rural road sub-projects (Ministry of Local Development, Government of Nepal, Kathmandu, 2010)

11. B.B. Deoja, Sustainable approaches to the construction of roads and other infrastructure in the Hindu Kush-Himalayas, ICIMOD, Occasional Paper No. 24 (1994)

12  Transport Research Laboratory (TRL), Principles of low cost road engineering in mountainous regions, with special reference to the Nepal Himalaya, Overseas Road Note16 (1997)

13  H.R Shrestha, Road vs. Hill Environment: the trend of road construction in Nepal, transport in mountains, in *An International Workshop Kathmandu, Nepal* (2010)

14  Decentralised Rural Infrastructure and Livelihood Project (DRILP), Road Cross-section, standard drawings. DoLIDAR (Ministry of Local Development, Government of Nepal, Kathmandu, 2006)

# Chapter 5
# Covering-Based Rural Road Networks

## 5.1 Introduction

Policymakers have recognized that building rural roads is a top priority for both rural populations and local governments in developing countries. A systematic planning approach is needed to create a network of rural roads. There are a number of transport network planning models available today, but they are primarily created with urban transport networks in mind, with the goal of reducing energy use, pollution, and traffic congestion. In contrast to urban transportation, rural transportation primarily focuses on ensuring that basic services are accessible to rural settlements. Local governments typically lack adequate funding as well as human resources and expertise. Therefore, complicated models are impractical in rural and hilly areas. As a result, the construction of rural roads in hilly regions of developing countries has mostly relied on spontaneous judgment and intuition. Lack of funding for developing rural infrastructure, including the road network, is one of the many challenges faced by the local government. Without suitable planning models, one of the most difficult problems for decision-makers in rural areas is how to make the most out of the available finite resources. As a result, a straightforward and efficient planning methodology for rural road networks is required.

In the context of Himalayan region, rugged topography consists of many streams and rivers that cut through the landscapes, rendering much of the country out of reach. Only the largest towns and cities—the majority of which are located in the southern plain regions—are connected by the existing highway system. However, a sizable portion of the population lives in the hilly and mountainous areas, where porters or pack animals are used for transporting goods. The inefficiency of transporting goods, even within plain areas, stems from the absence of road connections that are "all-weather" which connect road-accessible points to hinterlands. One of the primary causes of underdeveloped road networks in developing countries within the Himalayas region is the difficulty in geographical accessibility due to the challenging terrain.

J. Shrestha, *Rural Road Development in Developing Countries*,
SpringerBriefs in Applied Sciences and Technology,
https://doi.org/10.1007/978-981-96-2012-8_5

Additionally, due to the limited government funding, local resources must be mobilized for the development and upkeep of transportation infrastructure. However, transport network planning and prioritization are generally beyond the capabilities of a local government. To enable the creation of sustainable rural road networks, it is therefore required to develop and adopt an appropriate methodology as well as auxiliary instruments and procedures.

The transportation infrastructure requires significant investment because of the challenging topography in the hilly regions. Because of this as well as the limited financial resources that local governments have, a methodology that meets the minimum accessibility requirements while utilizing the resources at hand is necessary. Therefore, this chapter examines a straightforward and practical model for creating rural road networks in rural hilly areas to increase accessibility.

## 5.2  Rural Road Planning Models

Over the years, several rural road planning techniques have been developed by various authors and agencies. The following are some ideas and techniques that were proposed for planning rural road networks. This has been discussed in Chap. 3. Some of the key features of the previous models are briefly stated as follows:

The UNCHS [1] guideline is one of the earliest works on rural road network development techniques. In this guideline, the spanning tree concept is used, which has been extensively used in road network planning. A WSP decision model was introduced by Oudheusden and Khan in 1987 [2], which made it possible to optimize all-weather accessibility to a local market or a province road network under budgetary constraints. The rural road network model put forth by [3] minimizes the total cost of construction and transportation while connecting every village to the surrounding road network. To optimize the rural road network, [4] advocated the use of SST. To increase travel costs (person-km), the method produces a series of spanning tree networks, each of which is an improvement over the previous SST network. Additionally, [5] found that 93% of rural travel concludes at the next market center or educational facility. They also suggested the least expensive road links from each hamlet to these adjacent market centers and learning institutions.

The participatory district-level rural road network was proposed by [6] using the district accessibility criteria, nodal points, existing roadways and trails, local demand, and geographic features as a basis. It can be considered participatory in that it incorporates the community members at every stage of the decision-making process. Shrestha [7] developed a settlement interaction-based rural road network model by fusing the gravity model and the centrality index, taking into account the current rural transportation infrastructures as the basis. He subsequently discovered that the model could cut the length of the district-level roads from the 440 km suggested by the previous transport plan to 164 km, a decrease of over 62%, without significantly lowering the accessibility level.

For the objective of planning rural road networks, an interaction model based on settlements was proposed by [8]. The rural road networking in the model is based on the hub locations and the already-existing roads, tracks, and trails. This method enables the discovery of missing linkages based on current and potential nodal sites, existing links, and the assessment of the transport demand by developing an association with the centrality index and interaction intensity. Furthermore, the IRAP technique prioritizes road links according to settlement-based data by using the AI concept [9].

Typically, road planning exercises require the time-consuming and expensive collection of vast amounts of data for the economic evaluation of roadway links; however, in rural road planning, the results may not be as noteworthy. According to [10], it is rarely justified to prioritize rural road links based on economic models. One key factor in the rural link's economic justification could be the level of traffic on the road connection. [11] discovered that it was challenging to estimate trip generation from rural villages and obtain traffic data from rural settlements. Additionally, in a rural road link, traffic volume can be extremely low—less than 25 vehicles per day [12]. As a result, several different strategies have been used to evaluate rural road connections in rural regions.

Prioritizing rural road links was suggested by certain models based on agriculture potentiality [1]. However, agricultural potential is typically low in rural areas, especially in hilly regions. Prioritizing links was based on market centers, according to other models [3, 8]. Numerous communities and public facilities may be excluded from road access when road links are prioritized using the market center approach. Thus, the market-centered strategy might not be the best one for rural regions.

According to [2], the majority of rural road links cannot be justified using traditional economic models like BCA, ENPV, and EIRR. To evaluate and develop the rural road network in rural regions, various authors have proposed alternative assessment tools. These tools prioritize the public services and settlements' accessibility and connectedness over their economic implications. Therefore, a model has been developed based on accessibility and connectivity [2–5, 8, 9, 13].

Despite the fact that these techniques were tailored for rural road networks, they are not entirely suitable for use in the rural areas of the hilly region. [6] asserts that the participative district-level rural roadway system does not ensure the creation of an effective rural road network that serves the majority of the nearby settlements and public services. Although nodal points must be defined to analyze settlement interaction using the Shrestha [7] technique, there is no foundation for doing so. The approach disregards the coverage of habitations and public amenities and is skewed toward the market centers. The decision to choose a rural road linkage over connecting rural village settlements was made primarily for economic reasons and is skewed to connect densely inhabited and prosperous areas.

The approach is more market-centered in its methodology. According to Shrestha [10], the majority of the connections to rural settlements are not economically justified and a lot of data must be collected using this method to calculate a nodal point's centrality index. In rural areas, which are essentially hilly, accessibility to service centers and settlements is more crucial than financial gains.

Certain models [3, 4, 13] suggest connecting every rural settlement. However, due to financial and technical limitations, it is not feasible to connect all settlements in a rural area. Rural areas are characterized by sparsely distributed communities and public assets, which is even more in mountainous regions. As a result, it is not practical or feasible to link all of the communities and public facilities via rural roads. As a result, connectivity and accessibility to rural areas should be defined differently. Even though all of the communities cannot be connected, the rural roadway network should be situated as close to the communities and facilities as is practical. As a result, all settlements within a reasonable travel distance of the network should be served by rural road networks. Introducing the idea of covering—which might encompass all settlements and public spaces within a given radius—could be one approach to solving the issue. Since the models mentioned above haven't addressed this issue, they're also insufficient to handle issues with the road networks in rural areas of hilly regions. There is a significant void in the literature. Therefore, an aspect of this research is to address the issue of rural roadways by attempting to correct the shortcomings in the models that are now in use.

The problem is primarily addressed for plain areas in the current models of rural road networks. The issues encountered in rural areas with hills, however, are distinct. The covering is the part that is most often ignored in all models and methodologies. The methods hardly ever take into account covering public facilities and rural settlements.

In a rural road network, it is necessary to appropriately define and locate nodal points to fix road alignment in rural areas. This, however, is one of the more challenging undertakings for engineers and planners. To cover as many settlements and public assets as possible, the nodal points must be found. Nevertheless, not a single technique covered above has provided a foundation for locating nodal points, even in uncomplicated environments. No models have previously addressed this kind of problem. Hence, [14] have proposed a model that suggested identification of nodal points, connection of the nodal points, and generation of rural road network optimally in the hilly regions. This model is claimed to be a simpler and more practical model for roadways in rural, hilly regions. To cover most rural communities and public assets within a specified service distance, this chapter proposes a covering-based model.

## 5.3   Use of Location Models in Rural Road Network Design

In the past few decades, facility location models have been the subject of much intense research [15]. Numerous studies have demonstrated the usefulness of the mathematical models of location analysis in the location decision-making process involving public facilities, even though their application may be difficult for practitioners in developing countries. Locating the ideal configuration for installing one or more facilities to meet population demand is the focus of location-allocation problems.

[16] also employed location-allocation models to determine the best locations for medical facilities in a Bangladeshi rural area. It takes into account the number of facilities to be located and the maximum allowable travel distance using an integer programming formulation. The same authors later conducted a thorough review of these models [17]. In the subsequent work, the goals and benefits of these models for the location of rural health facilities were extensively demonstrated and explored.

In Daskin and Dean's (2004) work, the location set covering model, the maximal covering location model, and the $p$-median model—three conventional facility location models—were thoroughly evaluated and considered as the key models for healthcare facility planning. Those models have multiple applications in the field of healthcare.

In developing countries, one of the most widely used methods for determining the location of rural health facilities is probably location analysis using location models. Nonetheless, it is uncommon to define rural roads using location models. In the past, location theory and transportation network design have been studied independently. More work has recently been carried out on developing models that incorporate facility location with network design.

In two closely connected studies, [18] proposed a methodology for the network design and described optimum and heuristic approaches for discovering solutions. By generalizing the traditional simple plant location problem, the integrated model of facility location and transportation design for the networks was utilized to analyze the transportation planning scenarios. Sensitivity analyses were carried out for various scenarios across the papers to monitor the performance of integrated model. Nonetheless, the model might be excessively intricate and impractical for rural hill regions, where financial resources and subject-matter expertise are typically limited.

There are also works of literature devoted to the theoretical analysis of the planning of rural road networks. Most of this literature concentrates on linkage development instead of tackling the definition of the entire network. A few of the studies look into network design taking facility location into account. Nevertheless, by limiting their testing to simulated networks, those studies only offer a limited application. [19] also concluded that there isn't much research on the interaction of the two areas. Rather than aiming for a sophisticated model, the concept of a simple location-allocation model can be utilized to determine the obligatory points (control points) in an intricate system of rural roads.

Determining the optimal location of a facility is similar to determining the necessary points on a rural road network to make sure that every rural settlement is reached within a specified journey time. Therefore, one can approach the task of identifying mandatory points as a location analysis problem. The rural road network can be created by linking the designated mandatory points in the best possible way. Following this, the issue is handled as a location analysis issue, for which the sections that follow provide a method of resolution.

## 5.4 Proposed Rural Road Network Method

There are two stages to the proposed model. The initial phase of the method is to identify the nodal points, or mandatory locations, in the network of rural roads. The model in the second step defines the rural road network connecting the nodal points. A detailed description of the model can be found in the following subsections.

### 5.4.1 Identification of Nodal Points

The majority of the villages in rural hills lack access to public services and facilities within a reasonable travel time, and they are mostly unpopulated. Thus, it is crucial to identify one or more nodes that can reach most of the facilities within the specified ranges. As a result, village settlements must be grouped. Finding a suitable geographic location using political boundaries is one method of classifying the settlements. Fixing the location is challenging, though, because of the rough, hilly terrain. Therefore, a suitable technique is required to ascertain the position of these points.

Only a few of the dispersed villages on hill slopes are linked by rural roads; most are connected by foot trails. The distance is significantly larger than the Euclidian distance due to the trails' criss-crossing of the hill slopes, making the use of these distances impracticable. The villages are accessible via these networks of trails. Examining the path networks to find a central nodal village among all the settlements inside the allocated boundary is therefore beneficial from an accessibility standpoint. The lowest tier of the government can be taken as a basis of the political boundary. For an example, the ward boundaries may serve as the political boundary in the case of Nepal, the wards are the lowest tier of government. Further, the boundary can be divided into smaller regions. It is possible to create a transportation network inside a ward from which a distance matrix containing the real walking distances between each settlement and village can be derived. The shortest path matrix can then be obtained by applying the Floyd–Warshall algorithm [20] to determine the shortest distance between each node and the other nodes. Next, the central village of the political boundary can be identified and utilized as the nodal point. One or more nodes can be added if a single node is unable to cover the entire ward settlement within the allotted distance. The nodal points of the other regions can be located in a similar manner. We can make use of a facility allocation model for this. The location problem can be considered as nodal point location problems. The villages that encompass the majority of the remaining villages are the nodal point(s) that have been identified.

In location theory, common covering problems are as follows: locating a fixed number of facilities to maximize the number of covered demands (the maximal covering problem; [21]),locating a fixed number of places to minimize the maximum distance between a demand point and the facility assigned to cover its demands (the

$p$-center problem; [19]),and determining the minimum number of facilities needed to cover all demand nodes (the set covering problem).

A compromise on covering every settlement is required because financial and spatial limitations prevent covering every settlement. As a result, efforts should be made to maximize the coverage of settlements. Therefore, it appears that the maximal covering problem is the most suitable model in this context for determining the nodal points.

According to [21], the traditional Maximal Covering Location Problem (MCLP) is stated as follows. A limited number of users and facilities exist at a limited number of sites. In a network, nodes are denoted by the weights associated with these sets and locations, with arcs showing the distances between nodes. To access services from the designated number of facilities, the maximum travel distance for the villagers is set. The goal is to maximize the coverage provided by the allotted facilities, which is defined as the objective function (Eq. 5.1) and constraints (Eq. 5.2 to Eq. 5.5).

The MCLP is to maximize:

$$\sum_{i=1}^{n} a_i y_i \tag{5.1}$$

Subject to:

$$\sum_{j \in N_i} x_j \geq y_i \; 1 \leq i \leq n, \tag{5.2}$$

$$\sum_{i=1}^{n} x_i = p, \tag{5.3}$$

$$x_j = 0, 1 \; \forall j, \tag{5.4}$$

$$y_j = 0, 1 \; \forall i, \tag{5.5}$$

where

$x_j$   1 if a facility is located at $j$; $x_j = 0$ otherwise.
$y_i$   1 if demand from i is covered by a facility; $y_i = 0$ otherwise.
$N_i$   $\{j | d_{ij} \leq S\}$ is the set of facilities which are eligible to provide cover to demand $i$.
$n$   Number of demand points.
$a_i$   Population of demand $i$.
$d_{ij}$   Shortest distance between $i$ and $j$.
$p$   Number of facilities to be located.
$S$   Maximum service distance.

The coverage of settlements is influenced by the maximum service distance. Determining the maximum service distance is a calculated risk and determining a facility's catchment area is necessary when planning to enhance access. Since each person's travel time and distance have an impact on their welfare, it is crucial to take into account the maximum distance that each citizen must travel to access the relevant facility to prevent disparities in public service accessibility. The aim of the model is minimizing total cost should not be the only one taken into account, as it will encourage the placement of facilities in populated areas, penalizing other remote, low-density areas in the hilly regions. It is preferable to keep the travel time and distance from any village to the facility center within a certain bound [17]. The determination of the maximum service distance is a crucial matter, though. [16] solved an MCLP to find community clinics in a second study, setting a maximum distance of 2 km between the facilities and villages. A 4 km maximum service distance was sufficient to cover 99% of the rural population. However, finding the covering distance is a crucial first step. Depending on the scenario being studied, this covering distance may change considerably.

### 5.4.2 Defining the Rural Road Network

The rural road network may be defined using the nodal points that have been obtained, as well as any tracks, planned roads, and existing rural roads. It is necessary to determine any potential connections among the nodal points. These connections are technically possible but are limited by engineering and spatial limitations. The uneven geography and topography, especially in the hilly regions, frequently prevents several connectivity alternatives from being viable. The establishment and consideration of all feasible alternative options for connecting the two nodes is necessary for a more thorough examination of the rural road network.

The basic network is made up of the rural road links that connect the nodal points. It is necessary to create a distance matrix linking each nodal point and connecting point in the network of rural roads. Next, each nodal point in the network can be connected using Prim's Algorithm [22] to create a Minimum Spanning Tree (MST). The MST network is the bare minimum of connectivity required to make a given coverage distance accessible. The majority of the communities and public spaces in the area under consideration are served by this MST, which also serves as the rural road network.

## 5.5    Application of the Concept to the Hilly Regions of Nepal

Gorkha district is used as a test site for the applicability of the method in designing the rural road network; Fig. 5.1 shows this area which covers 15 wards of the district.

The wards are located between the rivers Burigandaki in the east and Daraundi in the west, to the north of the district headquarters, as Fig. 5.1 illustrates. Serving as a hub for the distribution of commodities and amenities to the district's wards, the

**Fig. 5.1** Location of the ward center and rural roads' network in the study area

district headquarters is the primary hub for district services. The national highway network of the nation is connected to the district headquarters through a feeder road, which is a road of higher standard.

The wards have a total of 63,437 residents. There are 219 villages spread across the district's 15 wards, which cover 244.2 km². Table 5.1 displays the number of villages, population, and ward area.

Figure 5.1 displays the current ward centers and road network connections in the area. Since all of the network's links are made of earth, they are only usable during the dry seasons. Only dirt paths exist for a portion of the links, making them unsuitable for vehicles. The Floyd–Warshall algorithm was used to obtain the distance matrix, which was then used to analyze the situation. The MCLP model has been used to determine the nodal points of the 15 wards. The associated short-distance matrices

**Table 5.1** Coverage provided by the nodal points for a service distance of 4 km

| Wards | Number of villages | Population | Area (km²) | Coverage by node | |
|---|---|---|---|---|---|
| | | | | Number | % |
| Aarupokhari | 13 | 5465 | 23.94 | 7 | 53.85 |
| Asrang | 13 | 3880 | 16.00 | 13 | 100.00 |
| Baguwa | 11 | 2246 | 6.34 | 11 | 100.00 |
| Borlang | 17 | 5383 | 29.95 | 13 | 76.47 |
| Bunkot | 24 | 7478 | 30.42 | 20 | 83.33 |
| Dhawa | 7 | 4040 | 16.05 | 6 | 85.71 |
| Finam | 18 | 3437 | 9.96 | 18 | 100.00 |
| Masel | 13 | 4408 | 14.96 | 11 | 84.62 |
| Nareshwor | 16 | 4501 | 13.13 | 14 | 87.50 |
| Panchkhuwa | 10 | 2422 | 9.69 | 10 | 100.00 |
| Pandrung | 13 | 3021 | 13.06 | 11 | 84.62 |
| Takukot | 16 | 4496 | 14.05 | 16 | 100.00 |
| Taku lakuribot | 16 | 2740 | 12.59 | 16 | 100.00 |
| Tandrang | 6 | 4928 | 15.83 | 5 | 83.33 |
| Taple | 26 | 4992 | 18.23 | 25 | 96.15 |
| Total | 219 | 63,437 | 244.20 | 196 | |

for each ward's network are calculated. The node which covers the maximum other nodes is considered as the nodal point.

Table 5.1 displays the coverage by nodal points in each ward. The ward name appears in the second column of the table, followed by population, area, and number of villages in the third, fourth, and fifth columns, respectively. The sixth column shows the total number of settlements covered by the nodal points; the seventh column shows the percentage of encompassed villages over all the villages in the wards. The nodal point's coverage ranges from 54 to 100% when the maximum service distance is set at 4 km. 90% coverage is obtained in the service distance on average.

As illustrated in Fig. 5.2, the solution to the covering problem yielded the nodal points for 15 wards. These nodal points serve as the foundation for defining the network's nodes and may be the location of the rural road network's mandatory points.

Finding the road connections to the nodal points of each of the 15 wards is the next step in the process. The majority of the nodal points in the region have been discovered to be connected by rural road links, as Fig. 5.2 illustrates. Taking into account technical and spatial restrictions, the distance matrix of the connecting and nodal points has been obtained. In the network, every nodal point must have at least one road link connecting it. Prim's algorithm is used to obtain the MST. Figure 5.3

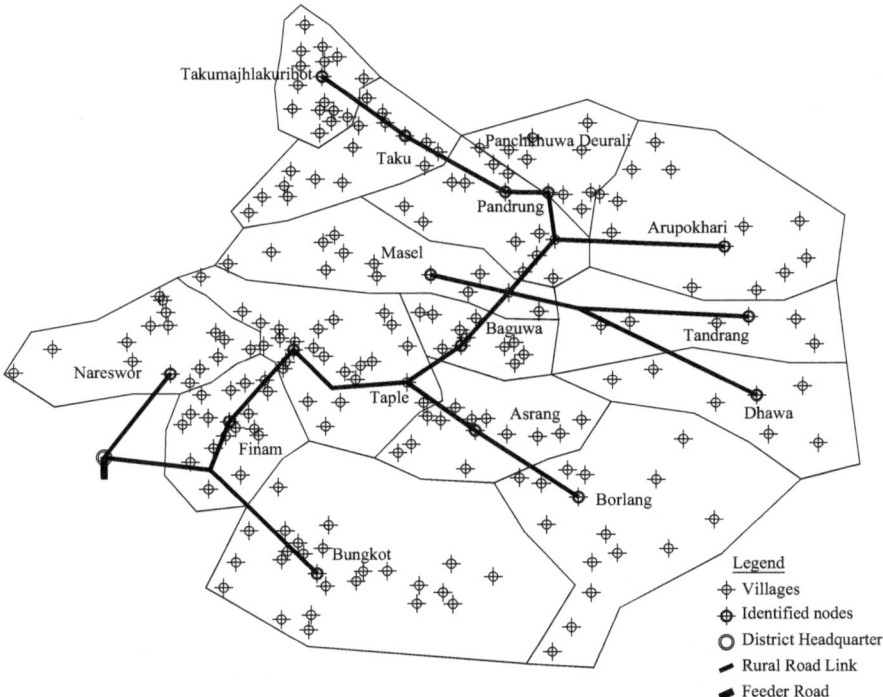

**Fig. 5.2**  Nodal villages obtained from the solution of the covering problem

displays the obtained MST, which is superimposed on Fig. 5.2. This is the minimum requirement to meet the criteria of a 4 km service distance of rural road connectivity.

The method's application in the 15 wards of the Gorkha district demonstrates how simple and practical it is for determining the proper mandatory points and the development of the rural road network. After the method's successful implementation in this area, three additional regions were studied [23]. The location of rural public facilities and the patterns of connection between rural road networks in hilly regions of Nepal have been analyzed using the study results, and the results are presented in the sections that follow.

## 5.6  Public Facility Locations in the Hilly Regions

In Nepal's hilly regions, settlements are widely dispersed. Even more dispersed are the public facilities for these settlements which are usually found at only one of these settlements. Schools, market centers, and health centers are typical public facilities in rural areas. These facilities are connected by foot trails and were typically situated sporadically. For the government to effectively provide services, it

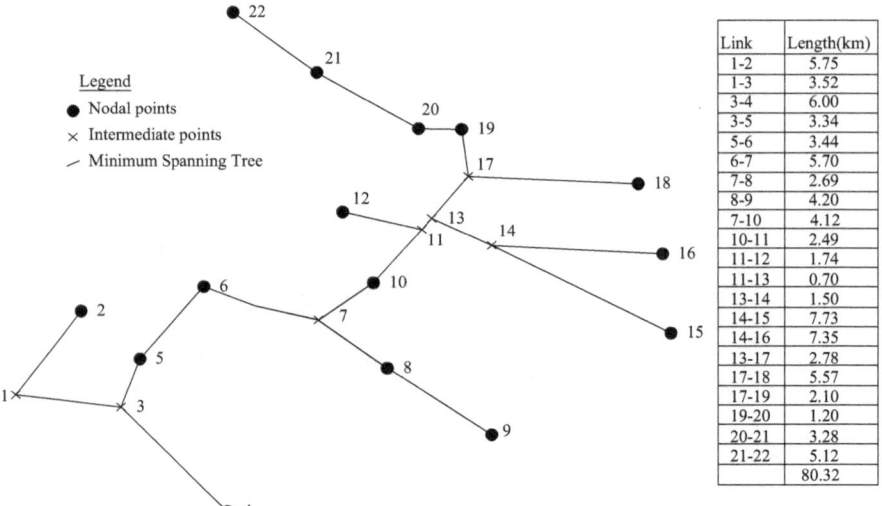

| Link | Length(km) |
|------|-----------|
| 1-2 | 5.75 |
| 1-3 | 3.52 |
| 3-4 | 6.00 |
| 3-5 | 3.34 |
| 5-6 | 3.44 |
| 6-7 | 5.70 |
| 7-8 | 2.69 |
| 8-9 | 4.20 |
| 7-10 | 4.12 |
| 10-11 | 2.49 |
| 11-12 | 1.74 |
| 11-13 | 0.70 |
| 13-14 | 1.50 |
| 14-15 | 7.73 |
| 14-16 | 7.35 |
| 13-17 | 2.78 |
| 17-18 | 5.57 |
| 17-19 | 2.10 |
| 19-20 | 1.20 |
| 20-21 | 3.28 |
| 21-22 | 5.12 |
|  | 80.32 |

**Fig. 5.3**   MST of the rural road network

is now crucial that these public facilities be connected by rural roads. However, due to financial, geographical, and environmental restrictions, it is not feasible to connect all of the facilities and the settlements in the country's hilly region with a rural roadway. Therefore, building a rural road network to make public facilities accessible from settlements provides a means of bringing facilities within the reach of rural communities. Covering a public facility's aspect is therefore more pertinent in a regional setting. Nevertheless, the appropriate covering distance might be a contentious political matter.

To better understand the circumstances in Nepal's mountainous regions, the coverage of public facilities and settlements has been studied across four regions [23]. In the study, the covering-based model was used to analyze the scenario. The objective of the study was to develop rural road network in four hilly regions of the two districts Gorkha and Lamjung in the country.

## 5.6.1   Covering of Settlements

The settlements were situated in four of the selected hilly areas of the Lamjung and Gorkha districts. The regions are made up of 63 wards with 1026 settlements. Figure 5.4 shows the impact of covering varying service distances (from 2 to 5 km at intervals of 0.5 km).

Figure 5.4 shows the rapid change in coverage from a 2 km to a 3.5 km service distance, after which it increases at smaller intervals. In this particular scenario, a 2 km service distance only covers 50% of the settlements. By increasing the coverage by

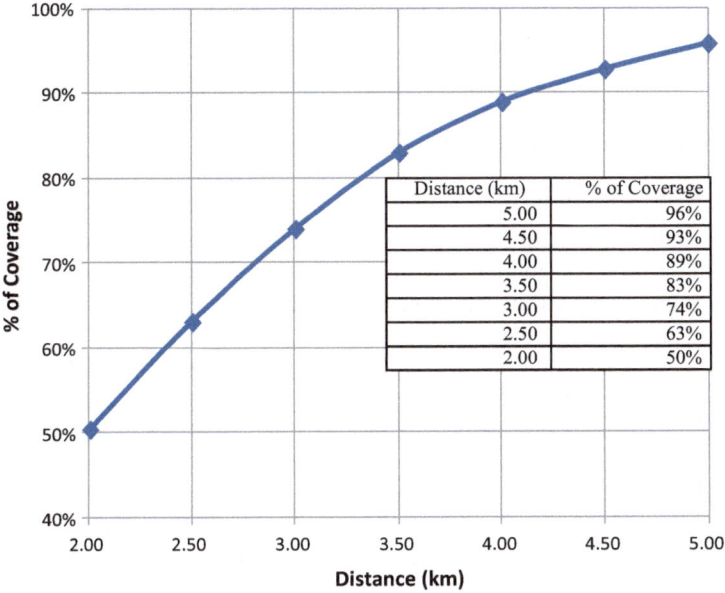

**Fig. 5.4** Effect of coverage in increase of service distance

24%, 74% of the communities are covered by the 3 km service distance. Additionally, service distances of 4 km and 5 km yielded the coverage of 89% and 96% of the settlements.

More settlements can be covered by reducing the service distance; however, this also makes it more difficult to provide service, which is undesirable because it reduces the nodal point's accessibility. To improve accessibility to the communities and public facilities, it is therefore desirable to add new nodal points as opposed to extending the service distance.

Over 80% of the villages are located within a 4 km walking distance, which suggests that the covering distance of 4 km can be established based on the results of the four case studies carried out in the hilly regions of the aforementioned districts. This value is in line with the 5 km/h average walking speed of humans. Alternatively, the walking distance of one hour is equivalent to the covering distance, which is reasonable in the hilly areas.

### 5.6.2 Covering of Public Facilities

In addition to the settlements, the covering issue also impacts service locations located in rural and hilly areas. As a result, the current public facilities—such as schools, health centers, and market centers—which have undergone analysis in the study area are as follows. Four regions were involved in the study: two in the Lamjung district and two in the Gorkha district with only 904 settlements and 56 wards included between the two districts [23].

Based on the coverage model, nodal locations have been determined, restricting the longest possible service distance to 4 km. The following sections include an analysis of the coverage status for public facilities.

**Covered Health Centers**

There is usually one health center per ward and these facilities are widely dispersed throughout the area. The model indicates that the nodal points' coverage of the health centers is found to be within a 4 km service distance. Data from health centers from 56 wards has been accumulated from the four study regions for the study. There are 60 health centers in the regions in total, 4 (7%) of which are located more than 4 km away from the nearest center. Additionally, 93% of all health centers are located within the set coverage distance.

**Covered Market Centers**

There are only a few market centers (18) in the study regions. Every market center has been identified to be protected by the nodal points.

**Covered Schools**

In every study region, schools are the most significant public facility. In a ward, schools are dispersed throughout various locations, much like the settlements. It has been observed that school coverage is less in comparison to other public facilities. In the four study regions, 328 (89%) of the schools were identified. 293 of the establishments are covered within the authorized coverage distance. 35 schools, or 11% of the total, are not inside the service region. This result is quite comparable to the nodal point coverage of 89% of the settlements (Fig. 5.4). To establish a rural road network in a given area and identify the nodal point (obligatory point) for a ward, it appears that school placements can be a useful guide. The map makes it simpler to locate schools. The settlements must, however, be grouped in a specific area. In rural, hilly areas, finding the centroid of a settlement's dispersed households is a persistent challenge due to their high dispersion.

**Overall Coverage**

Within a 4-kilometer radius, 90% of the facilities are covered, and 10% are found to be uncovered. Schools are found to be the most prevalent facilities in rural areas, so their distribution dominates the facilities.

When the nodal points are fixed per the aforementioned model, this study finds the market centers to be very well covered (100%). The majority of health centers (93%) in the regions are also included in the model with only the schools being the least covered (89%) and the most dispersed types of buildings. This further demonstrates that the majority of settlements and other public facilities in hilly regions cannot be included in the definition of a rural road network that only takes market centers into account. Nonetheless, the majority of the research on rural road planning place more emphasis on constructing rural roads to link market centers than on taking habitation and public infrastructure locations into account.

This study shows that school sites might be a useful tool for identifying the nodal points in rural areas that establish the rural road network. More settlements are found to be covered with more coverage of schools. Covering every community and facility in the area, though, can be an expensive undertaking. Consequently, the goal is to maximize the rural road's coverage of settlements and public facilities within a reasonable service distance. The coverage problem falls under budgetary and political circumstances. Nevertheless, the addition of more nodes to the wards with the least coverage can enhance the covering.

## 5.7  Linkage Pattern of Rural Roads in Hilly Regions

In hills of Nepal, numerous rural road networks were built in the past through community and public efforts. The majority of the networks are still in their infancy. The development of rural transportation routes is hindered by the hill's topography and geological characteristics, as observed in Nepal's Himalaya's area. In the hills, a rural road network layout has been formed by the existing network. These well-established layouts could serve as a basis for the development of a rural road network model for hilly regions. Hence, this section focuses on the developed pattern of rural road networks in hills. The locations of communities and public amenities influence how rural road development occurs in those areas. The schools, hospitals, and shopping malls are the facilities under consideration.

In the studied regions, the settlements are dispersed throughout the hill slopes. Settlements on the ridge slopes are typically heavily populated. Due to the steep and unstable hill slopes, fewer people are living in the valleys next to streams. Additionally, the streamlines are typically covered in forests and enclosed by unstable features like landslides and erosions. These regions are mostly arid because they are hard to cultivate. In Nepal's hilly regions, the stable land is either cultivated or inhabited. Therefore, stable areas are those where people have lived and farmed for a long time. Given that the upper hill slopes are less likely to experience instability issues and consist of denser settlements. The nodal points for the alignment of rural roads are also concentrated on hill slopes, usually along ridgelines. To link or provide accessibility to the settlements that are present, road alignment must follow the ridgeline as much as feasible.

The majority of the settlements are covered by the road that winds through stable slopes and hill ridges. Along with the villages, public amenities can be found near the ridgeline of the hill slopes. The analysis of public facility locations has found the same precedent (Sect. 5.6). In order to access the public establishments in the hills, it is therefore equally advantageous to locate road alignment along the slopes of the hills near to the ridgelines. This is an additional advantage of the rural road design that follows the ridge of the slope.

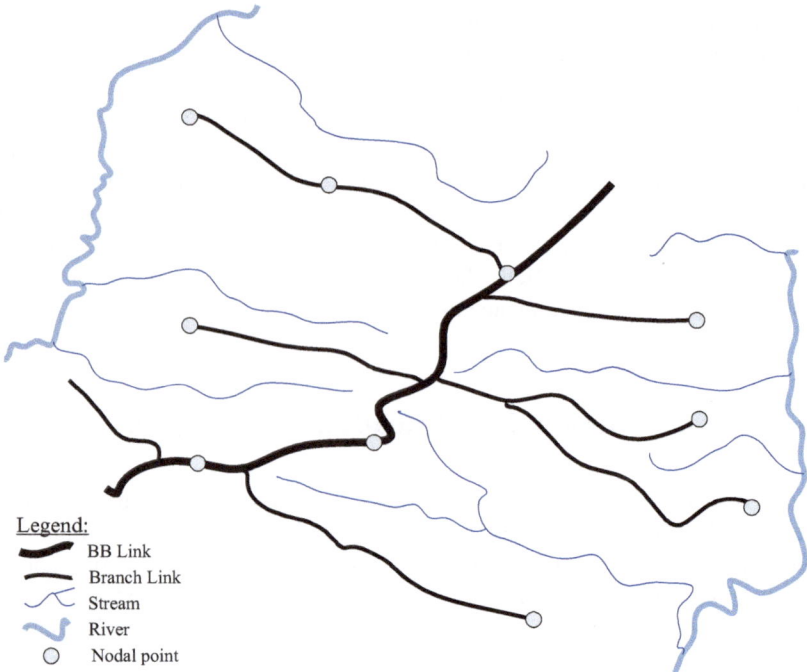

**Fig. 5.5**  Structure of backbone and branch road network in hilly region

## 5.7.1  Data Concerning the Rural Road Networks Under Study

In the four studies conducted on the rural road networks in hilly areas [23], every rural road connection to the nodal points that have been identified has been taken into consideration as an alternative link. As mentioned in Sect. 5.4, the covering-based model was used to identify the nodal points.

It is observed in each case that the main ridgeline that runs between two rivers on either side of the ridge is traversed by the backbone links. The secondary ridges that branch off of the main ridge are followed by the branch links. Identical formations can be found in the secondary ridges that lie between two streams where the streams are the tributaries of major rivers. Figure 5.5 illustrates this type of structure in the network.

## 5.7.2  Rural Road Network Formation

In the four case studies, the MST linkage lengths are compared with the existing rural road linkage lengths. It is found that the MST network linkages are significantly shorter than the original road links. The findings were derived from the case studies carried out in the districts of Lamjung and Gorkha, with a 4 km travel distance between the settlements and public facilities.

The overall MST road connectivity lengths for a 4 km service distance equal 43% of the total lengths of the original roads. This proves that the MST network—which accounts for less than 50% of the total road network in hilly/mountainous regions— can be focused on. A limited budget might be allotted to the key linkages (MST links), which essentially encompass public facilities and settlements. In addition, more studies can be done to establish the basic degree of connections and prioritize projects according to the availability of the financial resources.

It was discovered that the trail, tracks, and current roads connect the nodal points. Rural roads can take the place most of the trails/tracks. It is possible to upgrade the current tracks to upgrade to rural roads. Trails and tracks were used as a basis to construct the majority of the standard rural road in the past without any proper planning. The aforementioned study shows that the network of rural routes can be far shorter than it is now. This suggests that in the hilly regions of Nepal, the strategy works well to form network of rural roads. This can assist in determining the most essential rural road connections that should be developed, at least to a minimal standard. Therefore, it provides decision-makers with the options to pick which road links to enhance in order to better connect and cover settlements and public facilities. Furthermore, rather than allocating funds on a haphazard or politicized basis, the limited resource can be used more effectively in the crucial links. Investing in appropriate rural road connectivity can help to improve the quality of public services and accessibility in rural communities.

The major highway alignments are typically found along the major rivers. A rural road's route is drawn to the ridgeline from a location on a feeder road or from the highway at a river's side. One of the difficulties in figuring out the rural road alignment in a hilly region is locating the transitional alignment from the current road line at a river valley. Finding the transitional part of rural road alignments in hilly areas is one of the more challenging tasks. It is typically found on hill slopes. This may lead to drainage issues, erosion, and landslides on hillside slopes. Water drainage issues are more common in the valley slope areas. Therefore, choosing the transitional alignment for a hill rural road is an important step in the process. A plan should be implemented to ensure that the road's transitional length is both stable and minimal. The transitional portion carries the road head to the major ridgeline point where the backbone (main ridge road alignment) can form.

We can branch out a tree from the main ridge road alignment (backbone) to cover the public facilities and the rural hill settlements by connecting identified nodal points with road links. Based on these case studies, it is determined that the settlements and public facilities are sufficiently covered by alignments along the ridgelines.

In addition, in a different context, building roads through hills causes erosion and landslides on the surrounding natural slopes, as well as problems for valley-side agriculture and habitation. In many cases, this has led to the road itself sliding. The cost of damage could wind up being far greater than the cost of building the roads. The materials that were excavated, along with the forests, agricultural lands, and occasionally whole village settlements have also been washed away as a result of the damage. In the hills of Nepal, the majority of these roads have seriously harmed the environment. In this sense, planning, designing, and constructing rural roads in hilly areas presents environmental concerns in addition to social and technical ones that need to be carefully considered. A well-planned alignment of rural roads can minimize the issue by lowering the amount of cut that disturbs the surrounding ecosystem. Reducing the length of roads on hillslopes is one of the strategies to reduce the amount of cut material. In order to reduce environmental issues and the overall length of road links in rural road networks, it is important to optimize the road network in hilly areas. As previously mentioned, the MST reduces the length of the roads in a rural road network.

It is now possible to establish a basic procedure to develop a rural road network. It is stated as follows:

1. Determine and indicate on a map the major ridge line, or the backbone, which connects some of the nodal points.
2. On the same map, locate and label the branching (secondary ridge) lines in the hills so that the remaining nodal points can be joined to the backbone line.

One level of rural road network optimization is the MST, which offers a road network length that is substantially shorter than the current network while still providing better nodal point connectivity and coverage of public facilities and settlements. The covering-based method examined in this study has been found to be appropriate in defining rural road networks and nodal points in hilly areas.

## 5.8  Application to Other Rural Infrastructure Problems

The development and functioning of other infrastructures depend on the rural road, which is a fundamental component of rural infrastructures. The other vital physical infrastructures in rural areas include telecommunication, electricity, and water supply, all of which are rather expensive. In the same way as the size, density, distance, and purchasing power of these populations tend to increase investment costs, inappropriate facility site drives up the cost of these commodities in rural areas. There are certain limitations and mitigating strategies shared by the entire physical infrastructure issues mentioned above. The challenges include locating reservoirs for water, publicly accessible water tap stands, transformers and poles for electricity transmission, public phone booths, poles, and terminals. Additionally, the formation of the telecom wire network, the electricity distribution wire network, and the water distribution pipe network must be defined.

The following are some similarities between the issues of rural infrastructure:

- Coverage of settlements and public amenities.
- Determining the ideal position for nodal points.
- Establishing distribution networks.

Therefore, in order to accomplish the desired results with a lower cost and more efficient use of resources, the suggested covering-based method can be applied to problems with telecommunication, pipe water distribution, and electrical installation in rural areas. On these infrastructures, optimization problems incorporating factors like system reliability, cost, and/or efficiency can be developed.

## 5.9 Conclusions

This chapter proposes a model consisting of two steps to define a rural road network. The village nodes (also known as nodal points) that include the ward-containing settlements can be found using the covering model. In a network of rural roads, these nodes can be considered as the required locations. In the case of hilly regions of Nepal, the links to the nodes can create a basic network of rural roads. This technique offers a means of determining necessary points for rural road alignment, considering the coverage of settlements and public facilities in rural, hilly regions.

This model is helpful in determining how public facilities and settlements are covered by nodal points at various service distances. The public facilities and settlements are primarily located along hillside ridgelines. We can conclude that only roughly 50% of the settlements in the rural hill regions of the Lamjung and Gorkha districts are covered by the 2 km maximum service distance. The coverage considerably rises to 74% if we raise the maximum service distance to 3 km. In these cases, 89% of the settlements were covered within a 4 km service distance, and nearly 80% of the settlements and facilities were covered within a 3.5 km service distance. In hilly areas, a 4 km service distance can be considered an appropriate distance. By utilizing the MST, the refined rural road network length is considerably reduced to 43% for service distances of 4 km, ensuring improved coverage.

In the hilly regions of Nepal, the backbone and branch tree networks characterize the rural road network pattern. The branch tree network tracks the secondary ridgelines of the main ridgeline of hills, while the backbone links follow a main ridgeline. The pattern was discovered with the goal of finding the most cost-effective ways to link public facilities and rural hill settlements. In hilly areas, the suggested model may provide a feasible and realistic method for locating mandatory points in rural road networks. The model's applicability for the definition of the rural road network was confirmed by its implementation in Lamjung and Gorkha districts which lies in the hilly regions. In order to establish nodal points for developing rural water supply distribution networks, rural electrification grid extensions, and rural telecommunication network extensions, the covering-based rural road network model can also be used.

# References

1. UNCHS (United Nations Centre for Human Settlements), Guidelines for the Planning of Rural Settlements and Infrastructure: Road Networks (Nairobi, Kenya, 1985)
2. D.L. Oudheusden, L.R. Khan, Planning and development of rural road networks in developing countries. Eur. J. Oper. Res. **32**(3), 353–362 (1987)
3. A.K.Makarachi, H.T.Tillotson, Road planning in rural areas of developing countries. Europian J. Oper. Res. **53**, 279–287 (1991)
4. A. Kumar, H.T. Tilloston A, Planning model for rural roads in India, in *Procedings, Seminar on roads and road transport in rural areas* (Central Road Research Institute, New Delhi, India, 1985)
5. A. Kumar, P. Kumar, User friendly model for planning rural road. Transp. Res. Rec. **1652**, 31–39 (1999)
6. C.B. Shrestha, Manual for the preparation of a District Transportation Master Plan, Pilot Labour Based District Road Rehabilitation and Maintenance Project Butwal, Nepal (1997)
7. C.B. Shrestha, J.K. Routray, Application of settlement interaction based rural road network model in Nawalparasi district of Nepal, Transportation Research Board (2001), pp. 268–287
8. C.B. Shrestha, J.K. Routray, Application of settlement interaction based rural road network model in Nawalparasi district of Nepal, in *Technology transfer in road transportation in Africa: Arusha international conference centre*, Tanzania (2001) Conference proceedings (2002)
9. K. Dixon-Fyle, Accessibility planning and local development: the application possibilities of the IRAP methodology. RATP No. 2 (ILO-Geneva, 1998)
10. C.B. Shrestha, Developing a computer-aided methodology for district road network planning and prioritization in Nepal. Transp. Res. Board **3**, 157–174 (2003)
11. A. Athanasenas, Traffic simulation models for rural road network management. Transp. Res. Part E **33**(3), 233–243 (1997)
12. T. Airey, G. Taylor, Prioritization procedure for improvement of very low-volume roads. Transp. Res. Board Natl. Acad. **1652**, 175–180 (1999)
13. A.K. Singh, GIS Based rural road network planning for developing countries. J. Transp. Eng. (2010). https://doi.org/10.1061/(ASCE)TE.1943-5436.0000212
14. J.K. Shrestha, A. Benta, R. Lopes, C. Ferreira, N. Lopes, Covering-based rural road network methodology for hilly regions of developing countries: application in Nepal. J. Transp. Eng. **143**, 04016010 (2017). https://doi.org/10.1061/JTEPBS.0000013
15. Z. Drezner, H.W. Hamacher, *Facility location: Applications and theory* (Springer-Verlag, Berlin, 2004)
16. S. Rahman, D.K. Smith, Deployment of rural health facilities in a developing country. J. Oper. Res. Soc. **50**, 892–902 (1999)
17. S. Rahman, D.K. Smith, Use of location-allocation models in health service development planning in developing nations. Eur. J. Oper. Res. **123**, 437–452 (2000)
18. S. Melkote, M.S. Daskin, An integrated model of facility location and transportation network design. Transp. Res. Part A **35**, 515–538 (2001)
19. M.S. Daskin, S.H, Owen, Location models in transportation. in *Handbook of Transportation Science*, ed. R.W. Hall (Boston, MA: Springer US, 1999), pp. 311–360
20. R.W. Floyd, Algorithm 97: shortest path. Commun. ACM **5**(6), 345 (1962). https://doi.org/10.1145/367766.368168
21. Church R.L, ReVelle C, The maximal covering location problem. Pap. Reg. Sci. Assoc. **32**, 101–118 (1974)
22. R.C. Prim, Shortest connection networks and some generalizations. Bell Syst. Tech. J. **36**, 1389–1401 (1957)
23 J.K. Shrestha, *Rural Engineering Infrastructures Design and Public Facility Locations* (University of Aveiro, Portugal, 2013)

# Chapter 6
# Rural Road Network Optimization Models

## 6.1 Introduction

A significant population resides in rural, hilly, and mountainous areas of the Himalayan region, where public facilities are dispersed across numerous villages. Additionally, limited or inadequate road connectivity prevents many rural villagers from being integrated into the national road network. This lack of access to government services, as well as participation in economic and social activities, poses substantial challenges. Poor quality of life in these areas stems from several factors, including limited access to public transport linking people to essential services and the suboptimal placement of existing infrastructure relative to coverage needs. Quality issues also impact the developed rural infrastructure.

For example, due to challenging geography, only around 30% of Nepal is accessible by road. In hilly regions, over 39% of the population lacks access to all-weather roads within a four-hour walking distance [1]. Road networks are more established in the plains than in hilly areas, where road links between communities and public facilities remain underdeveloped. Consequently, expanding and upgrading these roads in hilly regions is essential to integrating rural residents and services with the national network.

Extensive research has been conducted on road networks and public facilities, though often in isolation. The layout of the network can restrict the reach of facilities; if the network is poorly designed, even optimally located facilities may be insufficiently accessible. Because the network interacts directly with facility regions, it is essential to consider facility location and network design together [2, 3]. Integrated rural infrastructure planning presents challenges, however, particularly in resource-limited developing regions where funds may be reallocated to other priorities, such as schools, hospitals, or new roads. Facility location models must account for these alternative resource uses and the potential for network enhancements, which may reduce average demand-weighted distance more effectively than simply adding new facilities. These models provide valuable support for decision-makers in allocating

© The Author(s), under exclusive license to Springer Nature Singapore Pte Ltd. 2025
J. Shrestha, *Rural Road Development in Developing Countries*,
SpringerBriefs in Applied Sciences and Technology,
https://doi.org/10.1007/978-981-96-2012-8_6

limited resources [2], and it is recognized that facility location and rural road networks are interdependent and should be studied systematically.

According to [4], beyond budget constraints, rural infrastructure development is hindered by the lack of suitable planning approaches. Models designed for urban settings often do not translate well to rural environments. Thus, the main aim of this chapter is to consider an integrated approach to planning public facility locations and rural roads to optimize budget allocation. This chapter presents models for rural road networks that incorporate the locations of public facilities, aiming to minimize total costs, including construction and maintenance expenses across various road surfaces—bituminous, gravel, and earthen. Public facilities in this context may include health centers, schools, and rural markets. The proposed model considers financial and geographical constraints in selecting optimal nodal points to deliver services and expand a rural road network.

## 6.2   Covering Aspects

Public amenities are widely dispersed across rural hills, where villages are often sparsely populated. This spread means that not all communities may have convenient access to essential services. However, ensuring adequate access to public facilities and services for all communities remains a priority. Therefore, identifying one or more central nodes that can serve multiple amenities, while considering distances, is essential. This central node is typically a village that serves a cluster of nearby villages. Using the geographic center of political boundaries is one method of grouping, although in hilly terrains, geographical centers may be less relevant. Furthermore, the rugged, mountainous landscape complicates locating such central points, emphasizing the need for effective methods to pinpoint these nodes.

In Chap. 5, we explored a coverage model suitable for placing public amenities in hilly regions. In these areas, only a few rural roads connect the scattered villages on steep slopes, with most connections made through foot trails. These footpaths serve as the accessibility network linking remote settlements and public facilities. Because the trails meander over the hills and cover far greater distances than a straight line, Euclidean distances are inappropriate for measuring access. The network of pathways can therefore guide the selection of a core village for locating a facility site, ensuring it is within reasonable distance from nearby villages and public services. Thus, assessing trail networks is essential for identifying a suitable nodal point (nodes being villages or facilities). A distance matrix can be constructed once actual distances between each village and facility are measured.

Public facilities may include market centers, health clinics, schools, and village development centers. The smallest political unit, such as a ward (as in Nepal), can be used to define the area of service. [5] suggest that evaluating the transportation system within a ward can help identify the village offering the best overall coverage and efficiency. This village can serve as the primary node for the ward. Additional nodes may be needed if a single node does not sufficiently cover all communities and facilities in the ward. For each ward, a distance matrix of communities and public services can be created to facilitate this process. Providing services in every village may be infeasible due to the small population sizes and significant financial costs involved. Decision-makers must weigh available resources against the extent of coverage that can be realistically achieved [6].

One strategy to reduce the number of facilities required is to relax the coverage criteria for demand points. Common coverage models include the set covering problem, which seeks the minimum number of facilities to cover all demand nodes, and the maximal covering problem, which locates a fixed number of facilities to maximize demand coverage [6]. In this context, the maximal coverage model is particularly relevant for locating nodal sites. A distance matrix is used to identify the shortest path between nodes, constructed using the Floyd–Warshall algorithm [7].

In Chap. 5, the mathematical formulation for coverage-based models is detailed. Limited sites with a high number of facilities and demand points are considered as nodes in a network, with the distances represented by weighted arcs. The goal is to specify the maximum travel distance that villagers should cover to access services from the chosen facilities, maximizing coverage with existing resources. The facilities are ideally situated in villages from which neighboring communities can be adequately supplied.

The distance residents must travel influences their choice to seek services from a specific facility. Hence, facility location must be carefully considered. Evaluating the optimal travel distance to facilities for residents is crucial to preventing significant disparities in service access, as individual travel distance (or time) affects quality of life. Although determining an appropriate travel distance can be politically sensitive, Chap. 5 recommends a travel distance of 4 km in hilly regions.

## 6.3  Rural Road Network Models

The basic network in the rural hills is formed by the rural road links connecting to nodal locations and facility nodes. The Prim's Algorithm [8] is used to connect each node in the network to produce a MST. As stated in Chap. 5, the MST network is the minimal degree of connectivity needed to cover public facilities in remote areas.

The network linkages can be considered possible links, potential tracks, damaged roads that can be maintained to become all-weather roads, or new links that could be used as candidate links for improving road surface options (earthen, gravel, or asphalt). The goal of the model that will be created is to have the lowest possible overall cost. The total cost includes all expenses associated with building, maintaining, and repairing roads.

The models are classified into two types. The first scenario involves new road construction, whereas the second involves improving existing road connectivity. In the second instance, some new link construction may be required and so can be included. The following two network models each cover one of the two scenarios.

A road network is thought to have residents and facilities on a number of village nodes, connected by road linkages as an undirected graph. A single surface in new construction and a variety of road surface options, including asphalt, gravel, and earthen, in link upgrades are taken into account in the mathematical formulation.

The model makes it possible to look at how best to allocate public resources while keeping overall costs as low as possible. We aim to construct the infrastructure to minimize overall transportation costs within the constraints of a small investment budget. Network nodal point connectivity is more crucial in new structures, though.

There are four rural road network concepts proposed, which will be looked at:

## Rural Road Network Model 1(RRNM-1): Upgrading Network

The majority of the network consists of current road connections that require surface-level upgrades. Few links can be added to an established network; those that are must be specified in terms of limitations. The model designates a road surface with possibilities for clay, gravel, or asphalt in order to minimize transportation costs within that budget level for the network. This versatile model can be used with both plain and hilly networks.

## Rural Road Network Model 2 (RRNM-2): New Network

All of the network's connections are newly established. This network could represent a universal network that works in plain and hilly areas.

## Rural Road Network Model 3 (RRNM-3): Upgrading Hill/Core Network

The model is one particular instance of model 1. According to the model, in hills, branch links should be enhanced after backbone links. A rural road network's core network can be identified and regarded as a backbone link in plain areas.

## Rural Road Network Model 4 (RRNM-4): New Hill/Core Network

This model is a particular example of Model 2. It is composed of branch links that follow subsidiary ridges and connect to the backbone links that travel along the main ridges of hills. The backbones are connected before any branch links are connected.

Below is a list of the models. A number of village nodes on a particular road network are considered to have residents. $G = (N, L)$ represents the network as an undirected graph where $N$ and $L$ represent the sets of village nodes and road links, respectively. The notations listed below are used.

$S$ is the set of road surface options $S = \{s1, s2, s3\}$ for earthen, gravel, and asphalt, respectively. $W_{ij}$ is the weight of the link $(i,j)$. $C_{ij}^s$ is the travel cost per unit flow over surface type $s \in S$ on link $(i,j)$. $d_{ij}$ is the distance from node $i$ to node $j$. $c_{ij}^s$ is the operating cost per unit flow of traveling over surface type s on link$(i,j)$. $O_{ij}^s$ is the operating cost on link $(i,j)$ over surface type $s \in S$, where $O_{ij}^s = d_{ij}c_{ij}^s$. $B$ is an available investment budget, and $I_{ij}^s$ is the cost of improving link $(i,j)$ with surface type $s$.

The decision variables in this model are: $x_{ij}^s=1$ if a link (i,j) is to be built with surface type $s$, 0 otherwise.

For new constructions, $S$ will have a single surface; hence, notation $s$ is not needed.

Budget constraints in this study are used to investigate different cases of decisions at different levels of budget.

The model can be built based on the Capacitated Facility Location/Network Design Problem (CFLNDP) [3], which tries to reduce overall population transportation costs subject to financial and spatial limitations, and can be restated as [9]:

RRNM-1

Minimize

$$z = \sum_{S=1}^{3} \sum_{(i,j)\in L} C_{ij}^{s} x_{ij}^{s} \tag{6.1}$$

The objective of the model function can be updated to account for operating expenses with weights assigned to the links [5]. Equation (6.1) can therefore be expressed as follows: Minimize

$$z = \sum_{S=1}^{3} \sum_{(i,j)\in L, i<j} W_{ij} O_{ij}^{s} x_{ij}^{s} \tag{6.2}$$

Subject to:

$$\sum_{S=1}^{3} \sum_{(i,j)\in L, i<j} I_{ij}^{s} x_{ij}^{s} \leq B \tag{6.3}$$

$$\sum_{S=1}^{3} x_{ij}^{s} = 1 \forall (i, j) \in L, \ i < j, \ \forall s \in S \tag{6.4}$$

$$x_{ij}^{s} \in \{0, 1\} \forall (i, j) \in L, \ i < j, \ \forall s \in S \tag{6.5}$$

Equation (6.3) indicates that the improvement/construction expenditure is constrained to an investment budget. Constraints (6.4) specify that just one link will be laid with a single type of surface. These constraints also ensure that all links are linked with one of the surface alternatives. The model is appropriate for problems involving link upgrades.

For new constructions, the objective function is different. The goal of Case 1 is to connect the links so that high-potential links are connected based on budget availability. The model can then be written as follows.

RRNM-2

Maximize

$$z = \sum_{(i,j)\in L, i<j} W_{ij} x_{ij} \tag{6.6}$$

Subject to:

$$\sum_{(i,j)\in L, i<j} I_{ij}x_{ij} \leq B \tag{6.7}$$

$$x_{ij} \leq 1 \ \forall (i,j) \in L, \ i < j \tag{6.8}$$

$$x_{ij} \in \{0, 1\} \forall (i,j) \in L, \ i < j \tag{6.9}$$

When the constraints are met, the models above can select any link in the network. These models apply to a broad range of situations.

In Chap. 5, a common network pattern in hilly areas has been identified. In hilly regions, a rural road network consists of backbone links and branch links that are connected to the former later. Therefore, before joining or upgrading the branch lines, the backbone linkages need to be improved. Plain areas' core network links can be identified and handled similarly to backbone links. The branch linkages decision variables can be further described as $x_{kl}$ for this purpose. As a result, the model will have more limitations.

Then, the following reconstruction of the model for the issues of upgrading the rural road network with branch/core and backbone network is possible:

RRNM-3
Minimize

$$z = \sum_{S=1}^{3} \sum_{(i,j)\in L, i<j} W_{ij}O_{ij}^{s}x_{ij}^{s} \tag{6.10}$$

Subject to:

$$\sum_{S=1}^{3} \sum_{(i,j)\in L, i<j} I_{ij}^{s}x_{ij}^{s} \leq B, \tag{6.11}$$

$$\sum_{S=1}^{3} x_{ij}^{s} = 1 \forall (i,j) \in L, \ i < j, \ \forall s \in S \tag{6.12}$$

$$\sum_{S=2}^{3} x_{ij}^{s} \geq x_{kl}^{s} \forall (i,j) \in L, (k,l) \in L, \ i < j, \ k < l, \ j \leq m, \ m < l, \ \forall s \in S \tag{6.13}$$

$$x_{kl}^{s} \in \{0, 1\} \forall (k,l) \in L, \ k < l \forall s \in S \tag{6.14}$$

Formulas 6.10, 6.11, and 6.12 are identical to Formulas 6.2, 6.3, and 6.4. Branch link nodes are numbered after backbone link nodes, where m is the greatest node number in the backbone network. Therefore, unless all of the backbone links are selected, Eq. 6.13 ensures that the secondary links cannot be selected.

Moreover, the following limitations would have been added to the earlier model if we had to take into account an intervention in branch connections only in cases where the backbone linkages had a greater surface level (gravel or asphalt). The branch links will never have an asphalt surface unless the backbone links are completely asphalt.

$$\sum_{S=t}^{3} x_{ij}^{s} \geq x_{kl}^{t} \forall (i,j) L, \ (k,l) \in L, \ i < j, \ k < l, \ j \leq m, \ m < l, \ \forall t \in S \qquad (6.15)$$

Similarly, the model can be updated and rewritten for new networks as follows:

RRNM-4

Maximize

$$z = \sum_{(i,j) \in L, i < j} W_{ij} x_{ij} \qquad (6.16)$$

Subject to:

$$\sum_{(i,j) \in L, i < j} I_{ij} x_{ij} \leq B \qquad (6.17)$$

$$x_{ij} \leq 1 \ \forall (i,j) \in L, \ i < j \qquad (6.18)$$

$$x_{ij} \geq x_{kl} \forall (i,j) \in L, \ (k,l) \in L, \ i < j, \ k < l, \ j \leq m, \ m < l \qquad (6.19)$$

$$x_{ij} \in \{0, 1\} \forall (i,j) \in L, \ i < j \qquad (6.20)$$

$$x_{kl} \in \{0, 1\} \forall (k,l) \in L, \ k < l \qquad (6.21)$$

## 6.4  Prioritization of Links

In developing countries, funds for rural road construction/improvement are typically scarce. As a result, the available resources must be used wisely. A prioritizing strategy is required for this. The rural roadway links in the network will be prioritized for implementation based on a realistic and practical criterion.

There are numerous strategies for prioritizing road connectivity. They are typically predicated on the economic gains from road connections. The net present value (NPV), discounted benefit–cost (B/C) ratio, and internal rate of return (IRR) are traditional economic feasibility metrics for highways. These traditional methods are applied on higher-quality, urban, and highway roads when it is possible to reasonably assess the economic return.

On the other hand, traffic volume is often low and economic activity is uncommon in rural areas. Less than 25 vehicles may drive on rural roads in mountainous regions each day [10]. The conventional economic indicator might not be suitable for rural roads as a result. Furthermore, it is challenging to assess the financial gains and advantages of rural roads. Road links in a network must therefore be categorized and prioritized using a different method.

### 6.4.1  Indicators for Rural Road Evaluation

Economic factors are generally considered when selecting road connections, with road traffic serving as the main indicator of a route's economic viability. However, districts rarely have traffic data for rural roads, as maintaining such data can be costly. Additionally, many rural hill road links may not be financially viable, as most see fewer than 25 vehicles per day [10]. From a social perspective, however, connecting public facilities and rural communities is essential to ensure minimum accessibility for goods and services.

In this context, a practical approach to prioritizing roadwork on low- and extremely low-volume routes in rural areas of developing countries is relevant [10]. This approach relies on using readily available data to forecast traffic on upgraded roads and considers the need to prioritize the opening of routes that may currently be inaccessible to vehicles.

Two scenarios are addressed: one for operational (passable) roads and one for impassable roads, which are further divided into two categories. The first category involves new construction, while the second pertains to roads in poor condition. For inaccessible roads, links with the lowest cost per capita are prioritized. However, a key challenge is quantifying the number of trips on these links; examples of potential trips include those for district, agricultural, and fishery purposes. Airey and Taylor [10] provide indicative data for district trips, yet estimating trips in rural hill contexts remains challenging.

An alternative approach is necessary to prioritize road links based mainly on social factors. In rural hill areas, the most significant social factor is the population served by each road link. Population growth drives traffic generation in rural road links, while other factors may have minimal impact in this setting. Therefore, a priority factor can be determined based on the population served by each link. Population weight can be related to trip generation, with links weighted when traffic and other relevant data are easily available.

The gravity model has been widely used in studies to examine traffic flow on road links. In social science terms, the gravity model represents a quantitative model of social interaction. According to [11], flow between two population centers is inversely related to distance or difficulty and directly related to the product of their populations, or their "attractive forces." That is,

$$F_{ij} = f\left(P_iP_j, \frac{1}{D_{ij}}\right) \tag{6.22}$$

More precisely, the most basic formulation is,

$$F_{ij} = k\frac{P_iP_j}{D_{ij}^b} \tag{6.23}$$

where $k$ and $b$ are empirically determined parameters. Population data can be gathered using census records, while distances between population centers are typically measured from maps. Jung et al. [12] studied the Korean highway system and applied the gravity model as a metaphor for physical gravity, formulating it as $T = f(P_1P_2/r^2)$, where $P$ represents population.

The objective of rural road design is to create a basic network that provides access to major centers, such as markets, clinics, schools, and other commercial, social, and welfare facilities in each village. These hubs are referred to as "market centers" and are expected to draw traffic from surrounding villages. Villages act as traffic generators, as they rely on rural roads to reach market centers, either directly or via main roads that link these centers.

Construction costs are generally proportional to the link lengths, making link length a practical basis for estimating construction costs [13]. Similarly, travel costs can be related to a measure known as "person-kilometers," which is calculated by multiplying the population at a village node by the distance from that node to its primary (or "root") node [13]. The following are the fundamental assumptions supporting the use of this factor:

- The number of trips generated by a village node is proportional to its population.
- The cost of travel is proportional to the distance traveled.

Thus, the person-kilometer factor, calculated as the product of population and distance traveled, is proportional to overall travel costs. This method is useful in rural areas, where data is often limited to basic information such as population figures and linear distances between villages. Notably, even when actual cost data is available, this approach remains applicable.

There are two main types of costs to consider. The first, construction costs, should include maintenance expenses distributed over the road's lifespan, which can typically be estimated with reasonable accuracy. However, the second type, travel costs, cannot be estimated with the same precision. To address this limitation, a model was developed that minimizes reliance on the exact values of travel costs. In relatively uniform topographic areas, it is assumed that construction costs are roughly proportional to the lengths of the required road links [14]. This model includes a weighted factor to account for different construction standards on certain links, although it maintains that costs are generally linked to road length. Travel expenses, on the other hand, are assumed to correlate with (i) the population connected by the link and (ii) the distance traveled via the link to reach a destination. Accordingly, travel costs are estimated using "person-kilometers," a measure defined as the product of the village's distance from the destination and the population it serves. This assumption applies best in areas with similar average income, community type, demographics, and agricultural practices, making the model suitable for relatively small, homogenous regions.

[15] proposed a model where the construction standards are determined by the population served by a link. In rural areas, where traffic data is typically scarce, population serves as a reliable substitute for traffic volume. Estimating traffic on rural roads requiring significant improvements and bridge construction is challenging. In such cases, road links can be prioritized by the population they serve and categorized by road surface type.

Two primary methods are generally used to prioritize rural roads: (a) sufficiency rating and (b) cost–benefit analysis. Evaluating the costs and benefits of a road in monetary terms is complex. Therefore, this model uses a simple metric—population served per unit of investment. As previously mentioned, rural road investment is viewed in terms of improving population access, so dividing the population served by a link by its construction cost offers a reasonable estimate of the road's benefit. Priority is given to the link that serves the largest population per unit investment. A rural road priority of a link can be calculated as follows:

Priority for a road link = Population served by the link/construction cost of the link.

DoLIDAR [5, 6] suggests using Cost Efficiency Analysis (CEA) to rank new transportation links. Points are assigned to criteria such as per capita cost and specific social considerations, including inclusiveness, based on their relative importance. Each road link is then given a score based on how well it meets these criteria. The total score for each link is the sum of the points allocated to each criterion, resulting in a ranked list of potential investments. Four key criteria are proposed to prioritize the development of new transport links, with indicators tailored to each road corridor's unique technical and socioeconomic data (see Table 6.1).

DoLIDAR [5, 6] considers population per unit cost a key factor in determining the priority of a road link. This parameter is calculated by dividing the total population served by the investment cost in hundred thousand rupees, representing the number of people per 100,000 Nepalese rupees (NRs). Additionally, the approach incorporates

**Table 6.1**  Scoring system for prioritization of new linkages [5, 6]

| Parameter | Scoring unit | Score |
|---|---|---|
| Population per unit cost | Population/investment cost in 100,000 | 55 |
| Cultivated land | Cultivated land/km | 15 |
| Population × walking hour | Population × walking hour/km | 20 |
| Total population of poor, *Dalits* and marginalized *Janjatis* | Population/km | 10 |

**Table 6.2** Scoring system for prioritization for upgrading and rehabilitation [5, 6]

| Criteria | Scoring unit | Score |
|---|---|---|
| Traffic unit | TU | 70 |
| Cost | Cost/km | 20 |
| Market/service center | Centrality index | 10 |
| Total | | 100 |

walking time to a road corridor, using population-distance or population-hour metrics to assess accessibility. It includes residents within the zone of influence (ZoI) area—defined as the area on either side of the planned road alignment—with a walking distance of up to two hours in flat terrain and four hours in hilly terrain. The road prioritization guidelines also consider agricultural land and areas within the ZoI that are home to impoverished and marginalized groups.

As shown in Table 6.2, the DoLIDAR [5, 6] manual provides specific indicators for prioritizing the upgrading and rehabilitation of existing rural roads.

Various types of vehicular and pedestrian traffic occupy the surface of rural transport links, each exerting different loads on the structure. Consequently, traffic volumes are measured using a standardized unit called the Transport Unit (TU) or Passenger Car Unit (PCU). This unit represents the traffic impact of a typical car, light van, jeep, or pickup traveling at 40 km/h. When another type of transport is compared to a standard car, its influence on traffic characteristics (such as headway, speed, and density) is referred to as the passenger car equivalent. Table 6.3 displays the traffic composition and related traffic coefficients.

The centrality index of the rural road network and the cost per kilometer for upgrades are key factors in ranking links. In developed areas, however, DoLIDAR [5, 6] favors a direct benefit approach, assessing rural roads through indicators like net present value, internal rate of return, and cost–benefit ratio.

### 6.4.2  Fixing Indicators

Prioritizing rural roads based on economic considerations is a challenging task to quantify. It is impracticable to prioritize rural roads in a network on the basis of economic factors in rural areas due to the challenges associated with determining the precise economic advantage. To gauge the value of a rural road connection, the population it serves is considered a useful proxy. This metric is frequently used in the literature to rank rural roads in order of importance.

Construction cost is another important consideration for evaluating and prioritizing rural roads. The cost of construction or upgrading should be justified by its benefits. On the other hand, calculating its advantages is challenging. Therefore, in several articles (e.g., [15]), links are prioritized based on the cost of population intervention. The link that serves the greatest number of people is regarded as a

**Table 6.3** Traffic unit [5, 6]

| Type of traffic | Transport unit (TU) |
|---|---|
| Cars, light vans, jeeps, and pick-ups | 1.0 |
| Light trucks up to 2.5 tons gross | 1.5 |
| Trucks upto10 tons gross | 3.0 |
| Trucks upto15 tons gross | 4.0 |
| 4W Tractor to wed trailers-standard | 3.0 |
| 2W Tractor to wed trailers-standard | 1.5 |
| Buses up to 40 passengers | 3.0 |
| Buses over 40 passengers | 4.0 |
| Bicycles | 0.5 |
| Rickshaws and tricycles carrying goods | 1.0 |
| Carts pulled/pushed by the human beings | 2.0 |
| Bullock carts with pneumatic tire wheels | 6.0 |
| Bullock carts with wooden wheels | 8.0 |
| Mule carts or horse-drawn carts | 6.0 |
| Pack animals and mules | 2.0 |
| Pedestrians walking on the link | 0.2 |
| Porters walking on the link | 0.4 |

prospective link in rural areas. As a result, the population that the link serves or the population per unit investment may be a helpful indicator for setting priorities.

Another factor to consider in the evaluation is the expense of travel for the rural population. This travel cost, however, is difficult to estimate because it is a time-consuming and costly work. To model the travel behavior of rural settlements, a huge amount of travel data is required, which is impractical. However, we must include the impact of travel costs when evaluating rural road linkages. The cost of transport may be more important than the cost of building a rural road connection.

The distance from the road head to the community and the village's population dictate the cost of transportation. The cost of travel increases with population density. In a similar vein, the cost of travel increases with the distance between the settlement and the road head. It is possible to measure the distance and get population statistics. However, estimating the cost of travel is challenging. Hence, indirect methods have been utilized in numerous papers/works to take into consideration for the influence of trip cost in the evaluation of rural road linkages. When assessing rural road links, one of the most important metrics is the person-kilometer, which has been used as an indirect indicator by numerous studies.

It is also possible to analyze traffic flow using the Gravity model. Estimating factors $k$ and $b$ in Eq. 6.23 presents the main difficulty. Certain research [12, 16] assume that $k$ is constant and that the value of $b$ is 2.

Most methodologies do not differentiate between indicators for new and existing linkages for upgrading or rehabilitation. DoLIDAR [5, 6], on the other hand, has addressed the prioritization task differently for new linkages and existing linkages that are to be considered for rehabilitation and upgrading. Many publications discovered that the travel cost was taken in the form of person-km and as a basic indicator to take into account the travel cost. This has been taken into consideration in the manual in the form of population hour. However, when compared to population per investment (55%), the weight attributed to the indicator was 20%. The remaining weights were 15% and 10% for cultivated land and 10% and 10% for marginalized poor populations, respectively.

The literature has used the same measures to assess both new and improved links. On the other hand, several indications are specified in the DoLIDAR [4, 5] manual for assessing upgrading links.

DoLIDAR [4, 5] focuses on the economic benefits of rural road restoration and upgrading based on traffic volume in road links. It has given 70% weight to the traffic indicator, 20% to the cost of upgrading, and 10% to the centrality index. It is, however, impractical to gather travel and traffic statistics in rural and hilly areas. Furthermore, although there is a lot less traffic in rural areas, it is still crucial to have good access to communities. Therefore, in most mountainous areas, indicators based on traffic units are inappropriate for determining the order of importance for upgrading and maintaining rural roads. The same indicators used to prioritize new linkages can be used to prioritize road upgrades.

Although DoLIDAR [4, 5] made an effort to account for significant characteristics, it is still challenging and unfeasible for rural hill sites. Data on traffic units, centrality indices, and farmed land may not be readily available, and even if they are, their significance may be questioned. Because of this, the criteria for ranking rural road links can be further streamlined by utilizing straightforward criteria like person-km, construction/maintenance cost, and population.

Access to rural communities and public services is one benefit of rural roads. The ease of access to market centers is the focal point of most initiatives. Still missing, though, is the indication that addresses the challenges associated with public facility coverage and settlement in rural areas.

## 6.5   Models' Application and Validation

As shown in Fig. 6.1, the applicability of the proposed models is evaluated in the rural road network presented in Chap. 5 in 15 wards of the hilly region of Gorkha district in Nepal.

The MST of the rural road network in 15 wards of the district is shown in Fig. 6.1. This is thought to be the minimum level of connectivity needed for the area because it links all nodal locations that are within a 4 km walking radius of the majority of communities and public amenities.

When adding a new link to an existing network or upgrading an existing one, the links must be prioritized. Here, the straightforward rural road prioritizing criteria

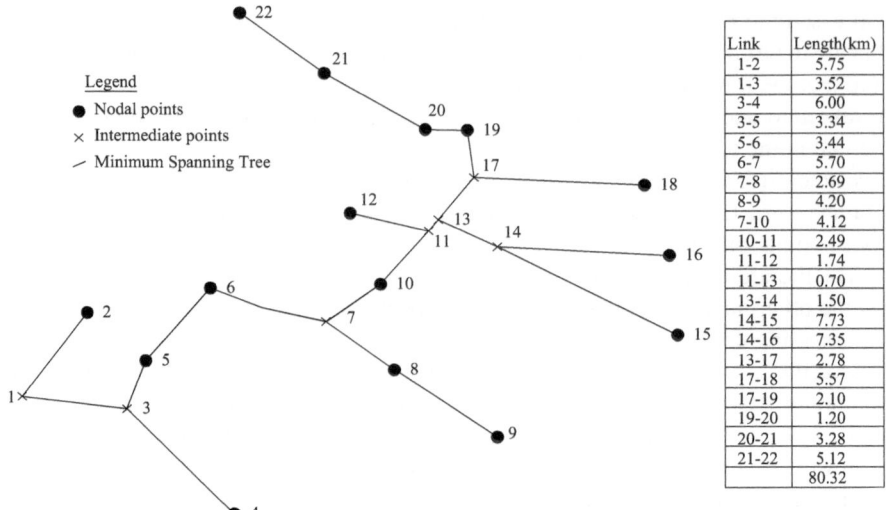

| Link | Length(km) |
|------|------------|
| 1-2 | 5.75 |
| 1-3 | 3.52 |
| 3-4 | 6.00 |
| 3-5 | 3.34 |
| 5-6 | 3.44 |
| 6-7 | 5.70 |
| 7-8 | 2.69 |
| 8-9 | 4.20 |
| 7-10 | 4.12 |
| 10-11 | 2.49 |
| 11-12 | 1.74 |
| 11-13 | 0.70 |
| 13-14 | 1.50 |
| 14-15 | 7.73 |
| 14-16 | 7.35 |
| 13-17 | 2.78 |
| 17-18 | 5.57 |
| 17-19 | 2.10 |
| 19-20 | 1.20 |
| 20-21 | 3.28 |
| 21-22 | 5.12 |
|       | 80.32 |

**Fig. 6.1**  Rural road network for model application

from Sect. 6.4 are applied. The weight of priority from four methods has been calculated separately as follows:

(i)   Population served by links (P1).
(ii)  Person-km (P2).
(iii) Population served/km (P3).
(iv)  Gravity flow model (P4).

The weight computed for each link—which serves as the number of trips for cost optimization—is displayed in Table 6.4. According to Tech Studio of Engineering (2011), the operating expenses per unit flow on asphalt, gravel, and earthen surfaces are NRs 36.79, NRs 45.64, and NRs 50.64, respectively (NRs are Nepalese Rupees, 1 Euro NRs 110). An estimate of NRs 5 million per kilometer is required for the upgrade from earthen surface level to gravel surface level, and NRs 10 million per kilometer is required for the upgrade from gravel surface level to asphalt surface level.

The mathematical models were solved using CPLEX 10.0's solver and MPL for Windows 4.2 as the modeling language. The analysis is carried out at various budget levels. The network node numbering is rearranged as shown in Fig. 6.2 for models RRNM-3 and RRNM-4. As shown in Fig. 6.2, these models feature backbone links (nodes 1 to 10).

In each model, the four methods of prioritization were applied. Table 6.5 (RRNM-1), Table 6.6 (RRNM-2), Table 6.7 (RRNM-3), and Table 6.8 (RRNM-4) demonstrate the intervention in the network link at various budget levels based on prioritization method P1 (population serviced by link). Based on the availability of funds, the decision-maker can select a set of links for intervention from these tables using the

**Table 6.4** Weight based on population, person-km, population per unit construction cost, and gravity flow model

| Links | Length | Population served | Person-km (cumulative) | Population served/km | Gravity flow | % of population served | % of cumulative p-km | % of population served/km | % of flow | Links after new BB scheme |
|---|---|---|---|---|---|---|---|---|---|---|
| 1–2 | 5.75 | 4501 | 25,881 | 783 | 3,509,997 | 7.10 | 1.9 | 2.02 | 51.74 | 1–11 |
| 1–3 | 3.52 | 58,936 | 1,364,905 | 16,743 | 6,784,469 | 92.90 | 100 | 43.23 | 00.00 | 1–2 |
| 3–4 | 6.00 | 7478 | 44,868 | 1246 | 2,127,380 | 11.79 | 3.29 | 3.22 | 31.36 | 2–12 |
| 3–5 | 3.34 | 51,458 | 1,112,582 | 15,407 | 4,657,089 | 81.12 | 81.51 | 39.78 | 68.64 | 2–3 |
| 5–6 | 3.44 | 48,021 | 940,712 | 13,960 | 1,856,578 | 75.70 | 68.92 | 36.04 | 27.37 | 3–4 |
| 6–7 | 5.70 | 43,029 | 775,520 | 7549 | 957,603 | 67.83 | 56.82 | 19.49 | 14.11 | 4–5 |
| 7–8 | 2.69 | 9263 | 47,526 | 3443 | 275,158 | 14.60 | 3.48 | 8.89 | 4.06 | 5–13 |
| 7–10 | 4.12 | 33,766 | 482,729 | 8196 | 512,915 | 53.23 | 35.37 | 21.16 | 7.56 | 5–6 |
| 8–9 | 4.20 | 5383 | 22,609 | 1282 | 1,184,016 | 8.49 | 1.66 | 3.31 | 17.45 | 13–14 |
| 10–11 | 2.49 | 31,520 | 343,613 | 12,659 | 996,213 | 49.69 | 25.17 | 32.68 | 14.68 | 6–7 |
| 11–12 | 1.74 | 4408 | 7,670 | 2533 | 553,313 | 6.95 | 0.56 | 6.54 | 8.16 | 7–15 |
| 11–13 | 0.70 | 27,112 | 257,458 | 38,731 | 442,900 | 42.74 | 18.86 | 100.00 | 6.53 | 7–8 |
| 13–14 | 1.50 | 8968 | 80,902 | 5979 | 135,176 | 14.14 | 5.93 | 15.44 | 1.99 | 8–16 |
| 13–17 | 2.78 | 18,144 | 157,578 | 6527 | 307,724 | 28.60 | 11.54 | 16.85 | 4.54 | 8–9 |
| 14–15 | 7.73 | 4040 | 67,450 | 523 | 58,823 | 6.37 | 4.94 | 1.35 | 0.87 | 16–17 |
| 14–16 | 7.35 | 4928 | 36,221 | 670 | 76,353 | 7.77 | 2.65 | 1.73 | 1.13 | 16–18 |
| 17–18 | 5.57 | 5465 | 30,440 | 981 | 92,170 | 8.61 | 2.23 | 2.53 | 1.36 | 9–19 |
| 17–19 | 2.10 | 12,679 | 76,697 | 6038 | 215,554 | 19.99 | 5.62 | 15.59 | 3.18 | 9–10 |
| 19–20 | 1.20 | 10,257 | 50,071 | 8548 | 5,695,718 | 16.17 | 3.67 | 22.07 | 83.95 | 10–20 |
| 20–21 | 3.28 | 7236 | 37,763 | 2206 | 1,379,806 | 11.41 | 2.77 | 5.70 | 0.34 | 20–21 |
| 21–22 | 5.12 | 2740 | 14,029 | 535 | 469,934 | 4.32 | 1.03 | 1.38 | 6.93 | 21–22 |

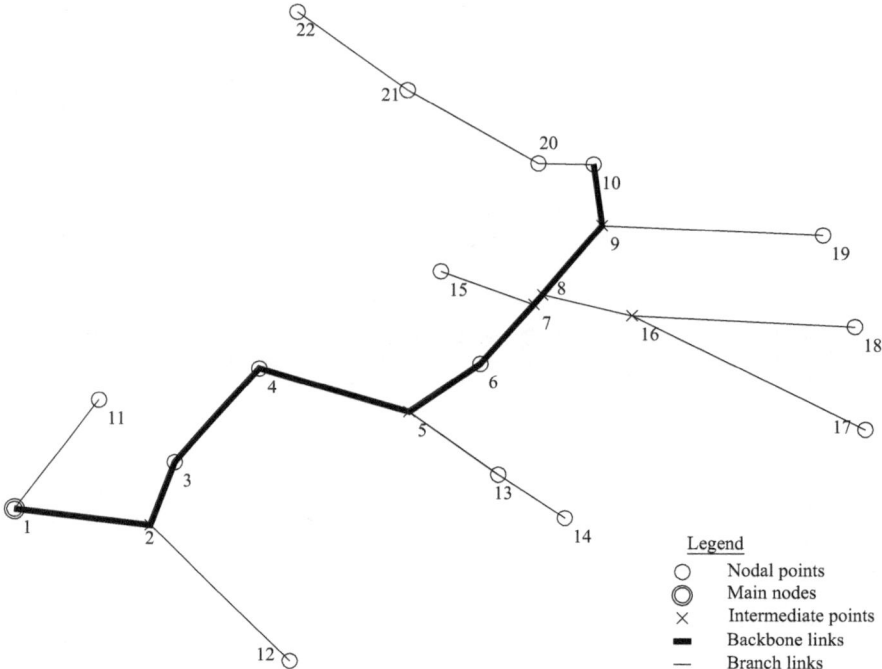

**Fig. 6.2** Node numbering scheme for model RRNM-3 and model RRNM-4 with backbone and branch links

rural road network model. Just four tables, for instance, are shown below. Similar constructions can be made using the output of these models (RRNM-1, RRNM-2, RRNM-3, and RRNM-4) using the other prioritization techniques, P2 (person-km), P3 (population served/km), and P4 (gravity flow model).

Figures 6.3, 6.4, 6.5, 6.6, 6.7, 6.8, 6.9 and 6.10 illustrate an optimized network intervention for different budget levels based different models.

The decision-maker may select a set of links for intervention based on the available budget. The recommended intervention may be found from RRNM-1 based on the four prioritization strategies independently, as illustrated in Figs. 6.3 and 6.4 for budget levels of NRs 400 million and 600 million, respectively.

In a similar manner, based on the four prioritization strategies, the output of the RRNM-2 is presented in Fig. 6.5 for budget-level NRs of 30 million and in Fig. 6.6 for budget-level NRs of 55 million. In this example, the network is taken to be brand-new.

The network is also analyzed using RRNM-3. Using the same four prioritization techniques, the budget in this instance ranges from NRs 100 million to NRs 800 million in order to validate the model's results. Figures 6.7 and 6.8 illustrate the suggested links with road surface for budget levels of NRs 400 million and NRs

**Table 6.5** Intervention in the network link at different budget levels based on P1 (RRNM-1)

| Links | Distance (km) | Budget (NRs) in millions | | | | | | | |
|---|---|---|---|---|---|---|---|---|---|
| | | 100 | 200 | 300 | 400 | 500 | 600 | 700 | 800 |
| 1–2 | 5.75 | | | | | | As | As | As |
| 1–3 | 3.52 | As | As | As | As | As | As | As | As |
| 3–4 | 6 | | | | As | As | As | As | As |
| 3–5 | 3.34 | | As | As | As | As | As | As | As |
| 5–6 | 3.44 | | As | As | As | As | As | As | As |
| 6–7 | 5.7 | As | As | As | As | As | As | As | As |
| 7–8 | 2.69 | | | | As | As | As | As | As |
| 7–10 | 4.12 | | | As | As | As | As | As | As |
| 8–9 | 4.2 | | | | | | As | | As |
| 10–11 | 2.49 | | As | As | As | As | As | As | As |
| 11–12 | 1.74 | | | | | Gr | Gr | | |
| 11–13 | 0.7 | As | As | As | As | As | As | As | As |
| 13–14 | 1.5 | | Gr | Gr | As | As | As | As | As |
| 13–17 | 2.78 | | | As | As | As | As | As | As |
| 14–15 | 7.73 | | | | | | | As | As |
| 14–16 | 7.35 | | | | | | | As | As |
| 17–18 | 5.57 | | | | | As | As | As | As |
| 17–19 | 2.1 | | | As | As | As | As | As | As |
| 19–20 | 1.2 | | | | Gr | As | As | As | As |
| 20–21 | 3.28 | | | | | As | As | As | As |
| 21–22 | 5.12 | | | | | | | | As |

*Note* As = Asphalt Gr = Gravel

600 million, respectively. The model has clearly chosen the backbone linkages first, followed by the branch links.

Finally, the network is analyzed using RRNM-4 for budget levels ranging from NRs 140 million to NRs 400 million. Figures 6.9 (budget 140 million) and 6.10 (budget 300 million) show the suggested links for the four intervention prioritization methods.

It is clear that when the budget is small (NRs 140 million), only backbone links are selected.

The majority of the chosen linkages have asphalt surfaces, as shown in Table 6.5. Given that gravel has a greater operating rate (NRs 45.54) than asphalt (NRs 36.79), this is the result of lower operating expenses in asphalt surfaces.

The suggested application of the models in the 15 wards of the Gorkha district demonstrates that they can be a workable and practical strategy for the creation of rural road networks and integrated facility locations in rural areas, especially in hilly terrain.

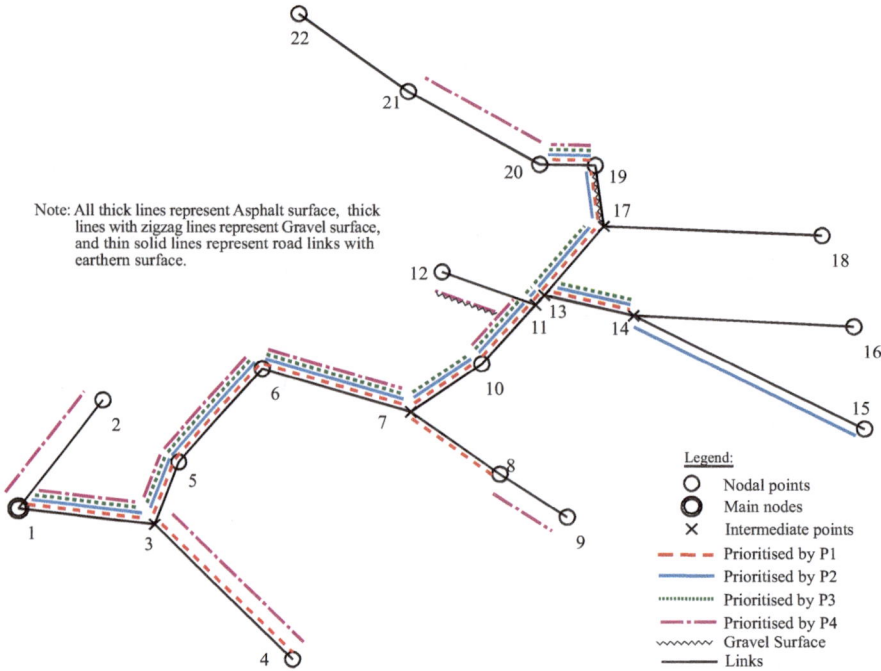

**Fig. 6.3** Optimal network intervention for a budget of NRs 400 million (RRNM-1)

The results of three approaches, P1, P2, and P3, are somehow comparable. The essential parameter in all of the techniques was population. The major issue in hilly places is accessibility, followed by others. As a result, the P2 technique—which takes into account both population and distance—might provide more practical guidelines for ranking new networks. The third method, P3, takes into account both population and building costs (by indirect distance calculation), and it could be a practical measure for improving rural roads in hilly regions.

## 6.6   Conclusions

Important variables for determining the priority of rural road links were found, including person-km, population per unit intervention cost, gravity flow, and the population served by the road link. For hilly areas, the same criteria apply for determining which rural road linkages should be prioritized for new construction and maintenance. The process of prioritizing becomes more difficult when there are too many indicators used, and the indicators may not have much of an impact on the mountainous region's rural parts. Therefore, for rural road prioritization, including a few key variables simplifies, makes sense, and might be adequate.

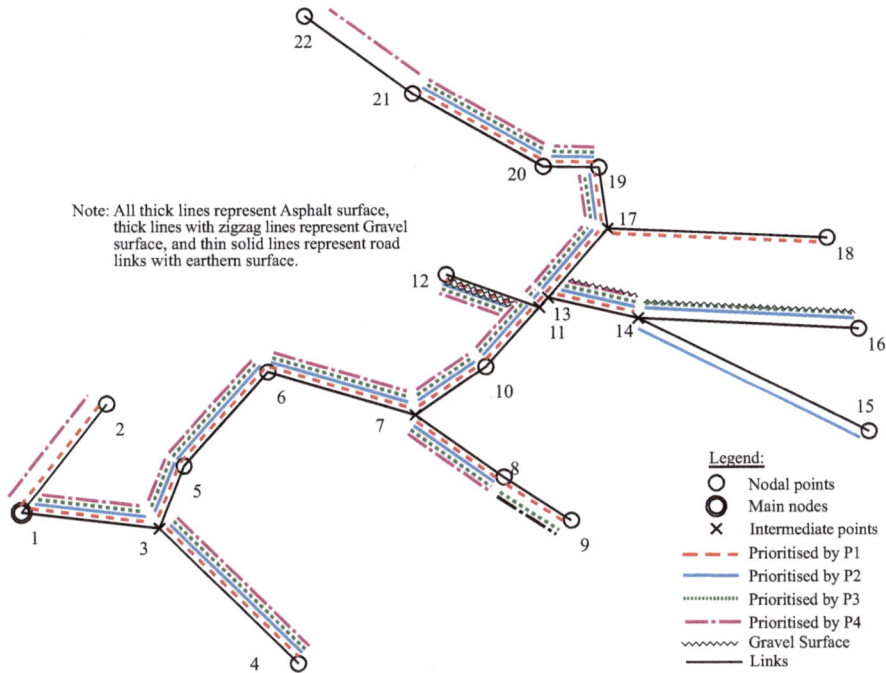

Note: All thick lines represent Asphalt surface, thick lines with zigzag lines represent Gravel surface, and thin solid lines represent road links with earthern surface.

Legend:
○ Nodal points
◉ Main nodes
× Intermediate points
– – – Prioritised by P1
——— Prioritised by P2
·············· Prioritised by P3
–·–·– Prioritised by P4
⌇⌇⌇⌇ Gravel Surface
——— Links

**Fig. 6.4** Optimal network intervention for a budget of NRs 600 million (RRNM-1)

This chapter examined the topic of creating a rural transport network to give rural inhabitants residing nearby better accessibility to public services and affordable road improvements. The suggested models optimize transportation costs on a rural road network with different types of road surfaces (earthen, gravel, or asphalt) by offering a portfolio of recommended linkages for road network enhancements and solutions for different budget levels. The suggested models may offer a more useful and realistic approach to researching and building rural road networks, especially in hilly terrain.

A two-step method for defining rural road networks in mountainous areas was discovered to be useful. It was proposed as a rural road network model based on coverage (as illustrated in Fig. 6.11). The nodal points that covered the settlements and public amenities within a region's specified limits were identified in the first step. These nodal locations were established as mandatory nodes in a rural road network. The model then constructed the linkages to the nodal locations in the second stage to form a basic road network in the given region. The MST was employed as a basic rural road network for the hilly regions, linking the selected nodal locations.

Within the networks in steep areas, a typical rural road network layout was discovered. The pattern was determined by the backbone and branch (BB) network. While the branch network follows subsidiary ridgelines (between two streams that are tributaries of the main rivers), the backbone network follows a key ridgeline (between two rivers).

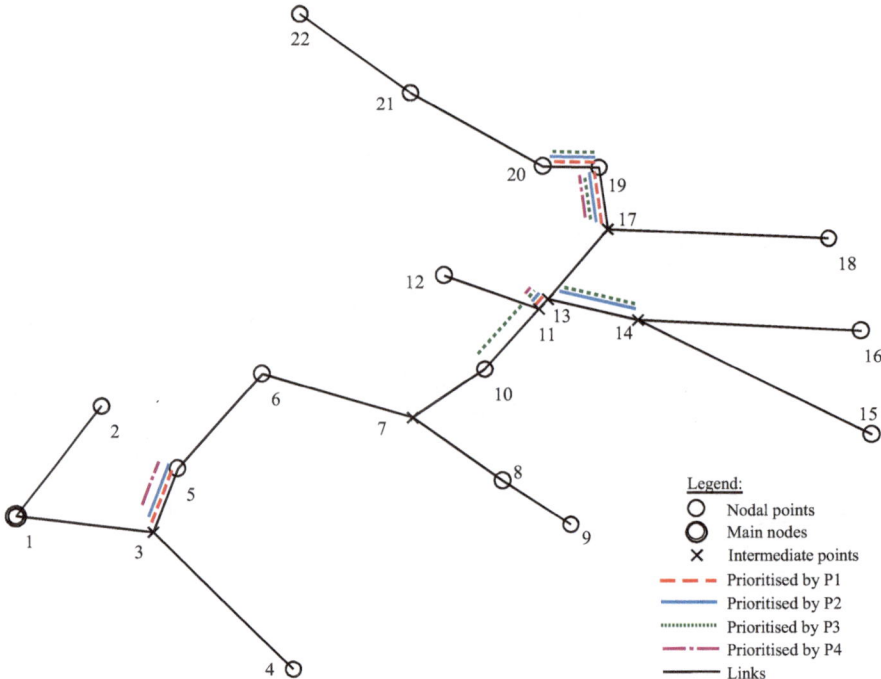

**Fig. 6.5** Optimal network intervention for a budget of NRs 30 million (RRNM-2)

The cases studied in four hilly regions of the two districts in Nepal, this model was utilized to analyze the coverage situation of villages and public facilities by nodal sites at different service distances. Only about half of the communities were found to be supplied within a 2 km maximum service radius. The coverage was increased to 74% when the maximum service distance was increased to 3 km. Within 4 km of the maximum service distance, 89% of the villages were serviced. The maximum service distance in the regions was designated as 4 km spanning service distance.

For maximum service distances of 4 km, the road network length was found to be significantly shorter (43% of the existing network length), ensuring better coverage of settlements and public facilities in the case study regions. This has been determined to be the minimum level of connectivity/coverage required.

Four models were identified as suitable for rural road networks and offered as rural road network choice models (shown in the second section of Fig. 6.11). Two of them (RRNM-1 and RRNM-2) are suitable for general-purpose (may be utilized in both flat and hilly road networks) models. The other two types (RRNM-3 and RRNM-4) are suitable for hilly locations where the BB network design is used. However, in plain regions, a core network within a network can also be defined, and the notion of BB network can be applied. For a new network with a single surface (earthen, gravel, or asphalt), the RRNM-2 and RRNM-4 models were proposed. For the improvement of existing rural road networks, the RRNM-1 and RRNM-3 models were developed.

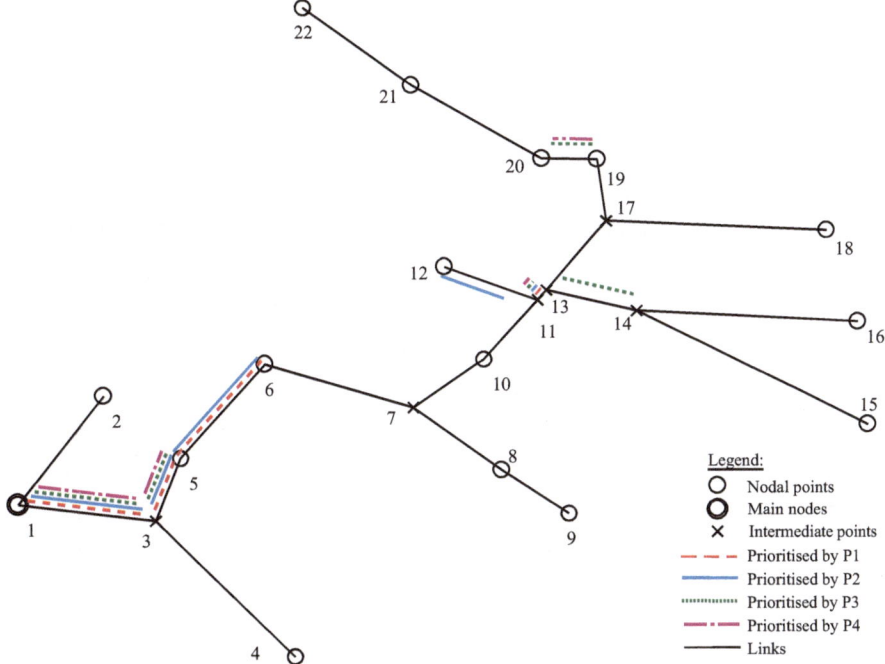

**Fig. 6.6** Optimal network intervention for a budget of NRs 55 million (RRNM-2)

The population served by a road link, person-km, population per unit intervention cost, and gravity flow model based on socioeconomic characteristics are all suggested as significant metrics for prioritizing rural road links.

The models supplied a portfolio of suggested linkages for road network enhancements, as well as solutions for various budget levels for optimizing transportation costs in a rural road network with various types of road surface (earthen, gravel, or asphalt) for upgrading.

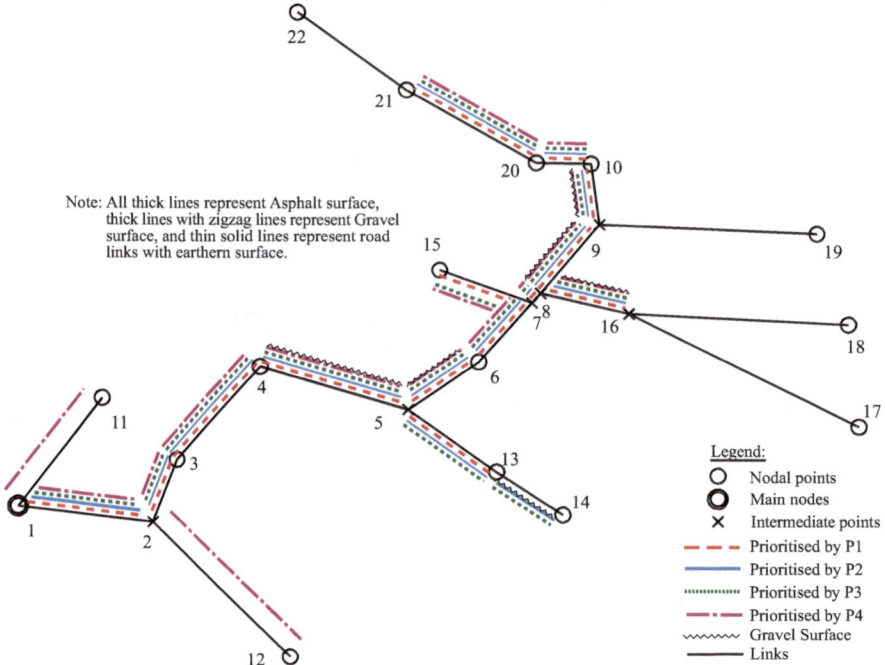

**Fig. 6.7** Optimal network intervention for a budget of NRs 400 million (RRNM-3)

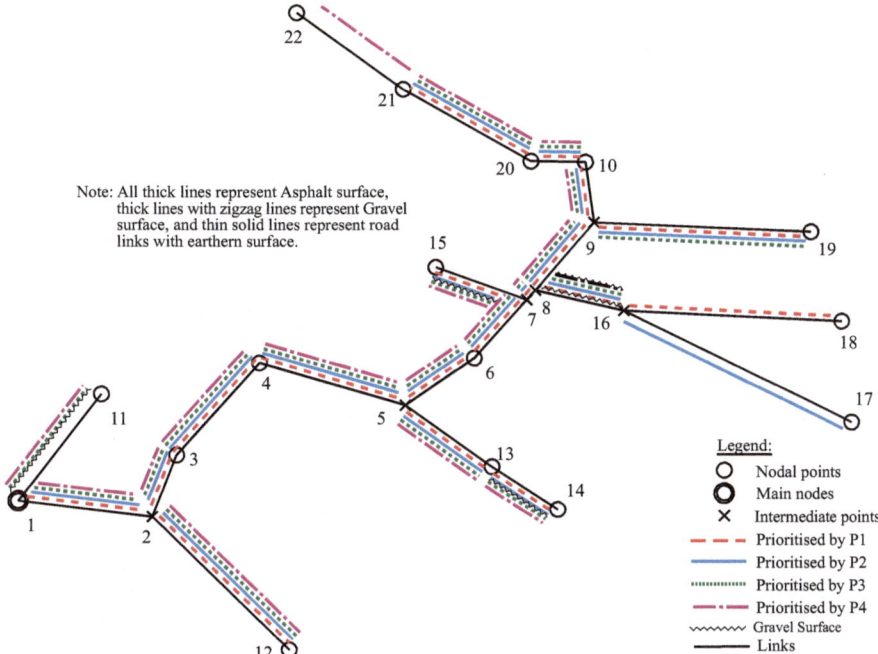

**Fig. 6.8** Optimal network intervention for a budget of NRs 600 million (RRNM-3)

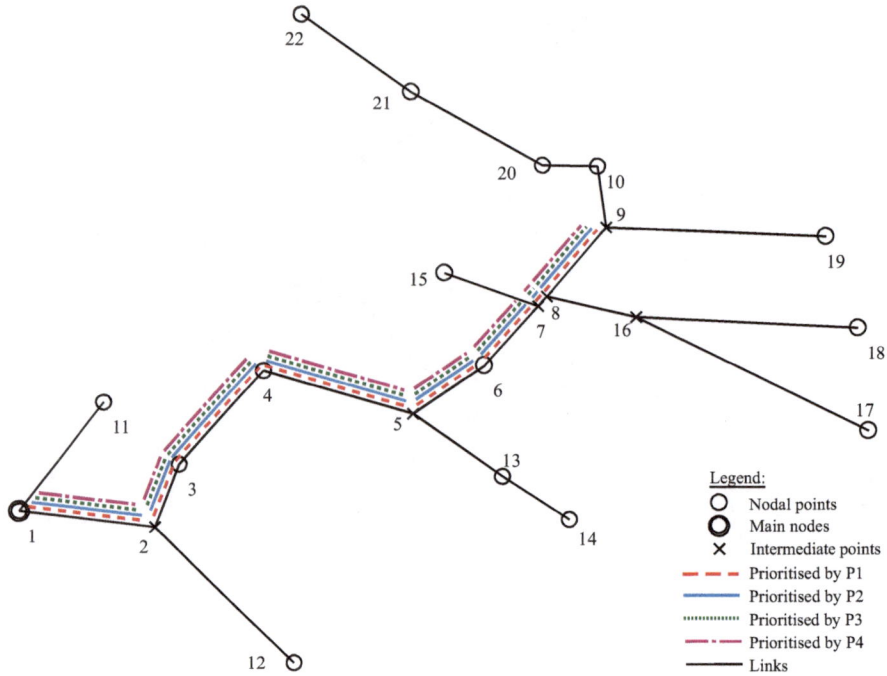

**Fig. 6.9** Optimal network intervention for a budget of NRs 140 million (RRNM-4)

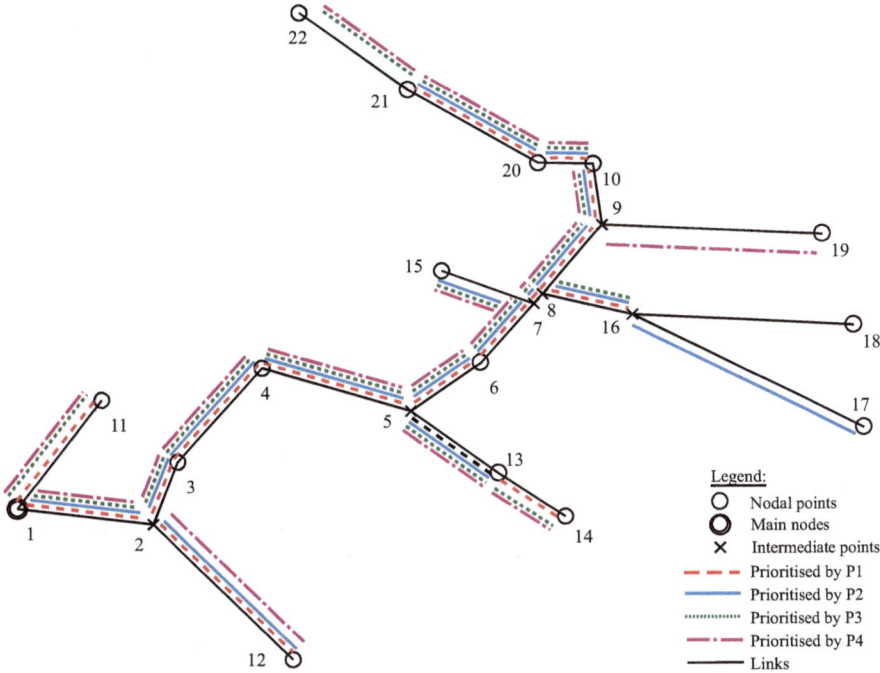

**Fig. 6.10** Optimal network intervention for a budget of NRs 300 million (RRNM-4)

**Table 6.6** Intervention in the network link at different budget levels based on P1 (RRNM-2)

| Links | Distance (km) | Budget (NRs) in millions | | | | | | | | | |
|---|---|---|---|---|---|---|---|---|---|---|---|
| | | 10 | 15 | 20 | 25 | 30 | 35 | 40 | 45 | 50 | 55 |
| 1–2 | 5.75 | | | | | | | | | | |
| 1–3 | 3.52 | | | Er | Er | Er | Er | Er | Er | Er | Er |
| 3–4 | 6 | | | | | | | | | | |
| 3–5 | 3.34 | | | | | | | Er | Er | Er | Er |
| 5–6 | 3.44 | | | | | | | | | Er | Er |
| 6–7 | 5.7 | | | | | | | | | | |
| 7–8 | 2.69 | | | | | | | | | | |
| 7–10 | 4.12 | | | | | | | | | | |
| 8–9 | 4.2 | | | | | | | | | | |
| 10–11 | 2.49 | | | | | | Er | | Er | | |
| 11–12 | 1.74 | | | | | | | | | | |
| 11–13 | 0.7 | Er | Er | Er | | Er | Er | Er | Er | Er | Er |
| 13–14 | 1.5 | | | | | | | | | | |
| 13–17 | 2.78 | | | | | | | | | | |
| 14–15 | 7.73 | | | | | | | | | | |
| 14–16 | 7.35 | | | | | | | | | | |
| 17–18 | 5.57 | | | | | | | | | | |
| 17–19 | 2.1 | | | | | | | | | | |
| 19–20 | 1.2 | Er | Er | | | Er | | | Er | | |
| 20–21 | 3.28 | | | | | | | | | | |
| 21–22 | 5.12 | | | | | | | | | | |

*Note* Er = Earthen

**Table 6.7** Intervention in the network link at different budget levels based on P1 (RRNM-3)

| Links | Distance (km) | Budget (NRs) in millions | | | | | | | |
|---|---|---|---|---|---|---|---|---|---|
| | | 100 | 200 | 300 | 400 | 500 | 600 | 700 | 800 |
| 1–2 | 5.75 | | As | As | As | As | As | As | As |
| 1–3 | 3.52 | As | As | As | As | As | As | As | As |
| 3–4 | 6 | As | As | As | As | As | As | As | As |
| 3–5 | 3.34 | | As | As | As | As | As | As | As |
| 5–6 | 3.44 | | As | As | As | As | As | As | As |
| 6–7 | 5.7 | As | As | As | As | As | As | As | As |
| 7–8 | 2.69 | As | As | As | As | As | As | As | As |
| 7–10 | 4.12 | | | As | As | As | As | As | As |
| 8–9 | 4.2 | | | Gr | As | As | As | As | As |
| 10–11 | 2.49 | | | | | | | As | As |
| 11–12 | 1.74 | | | | | As | As | As | As |
| 11–13 | 0.7 | | | | As | As | As | As | As |
| 13–14 | 1.5 | | | | | As | As | As | As |
| 13–17 | 2.78 | | | | As | As | As | As | As |
| 14–15 | 7.73 | | | Gr | As | As | As | As | As |
| 14–16 | 7.35 | | | | | | | | As |
| 17–18 | 5.57 | | | | | | Gr | As | As |
| 17–19 | 2.1 | | | | | | As | As | As |
| 19–20 | 1.2 | | | As | As | As | As | As | As |
| 20–21 | 3.28 | | | | As | As | As | As | As |
| 21–22 | 5.12 | | | | | | | | Gr |

*Note* As = Asphalt Gr = Gravel

**Table 6.8** Intervention in the network link at different budget levels based on P1 (RRNM-4)

| Links | Distance (km) | Budget (NRs) in millions | | | | | | | | | | |
|---|---|---|---|---|---|---|---|---|---|---|---|---|
| | | 140 | 160 | 180 | 200 | 220 | 240 | 260 | 280 | 300 | 320 | 340 |
| 1–2 | 5.75 | Er | Er | Er | Er | Er | Er | Er | Er | Er | Er | Er |
| 1–3 | 3.52 | Er | Er | Er | Er | Er | Er | Er | Er | Er | Er | Er |
| 3–4 | 6 | Er | Er | Er | Er | Er | Er | Er | Er | Er | Er | Er |
| 3–5 | 3.34 | Er | Er | Er | Er | Er | Er | Er | Er | Er | Er | Er |
| 5–6 | 3.44 | Er | Er | Er | Er | Er | Er | Er | Er | Er | Er | Er |
| 6–7 | 5.7 | Er | Er | Er | Er | Er | Er | Er | Er | Er | Er | Er |
| 7–8 | 2.69 | Er | Er | Er | Er | Er | Er | Er | Er | Er | Er | Er |
| 7–10 | 4.12 | Er | Er | Er | Er | Er | Er | Er | Er | Er | Er | Er |
| 8–9 | 4.2 | | Er | Er | Er | Er | Er | Er | Er | Er | Er | Er |
| 10–11 | 2.49 | | | | | | | | | Er | | |
| 11–12 | 1.74 | | | | | | Er | Er | Er | Er | Er | Er |
| 11–13 | 0.7 | | | | Er | Er | Er | Er | Er | Er | Er | Er |
| 13–14 | 1.5 | | Er | | | Er | Er | | Er | Er | Er | Er |
| 13–17 | 2.78 | | | Er | Er | Er | | Er | Er | | Er | Er |
| 14–15 | 7.73 | | Er | Er | Er | Er | Er | Er | Er | Er | Er | Er |
| 14–16 | 7.35 | | | | | | | | | | | |
| 17–18 | 5.57 | | | | | | | | | | Er | Er |
| 17–19 | 2.1 | | | | | | Er | | | | Er | Er |
| 19–20 | 1.2 | Er | Er | Er | Er | Er | Er | Er | Er | Er | Er | Er |
| 20–21 | 3.28 | | | | Er | Er | Er | Er | Er | Er | Er | Er |
| 21–22 | 5.12 | | | | | | | | | | | Er |

*Note* Er = Earthen

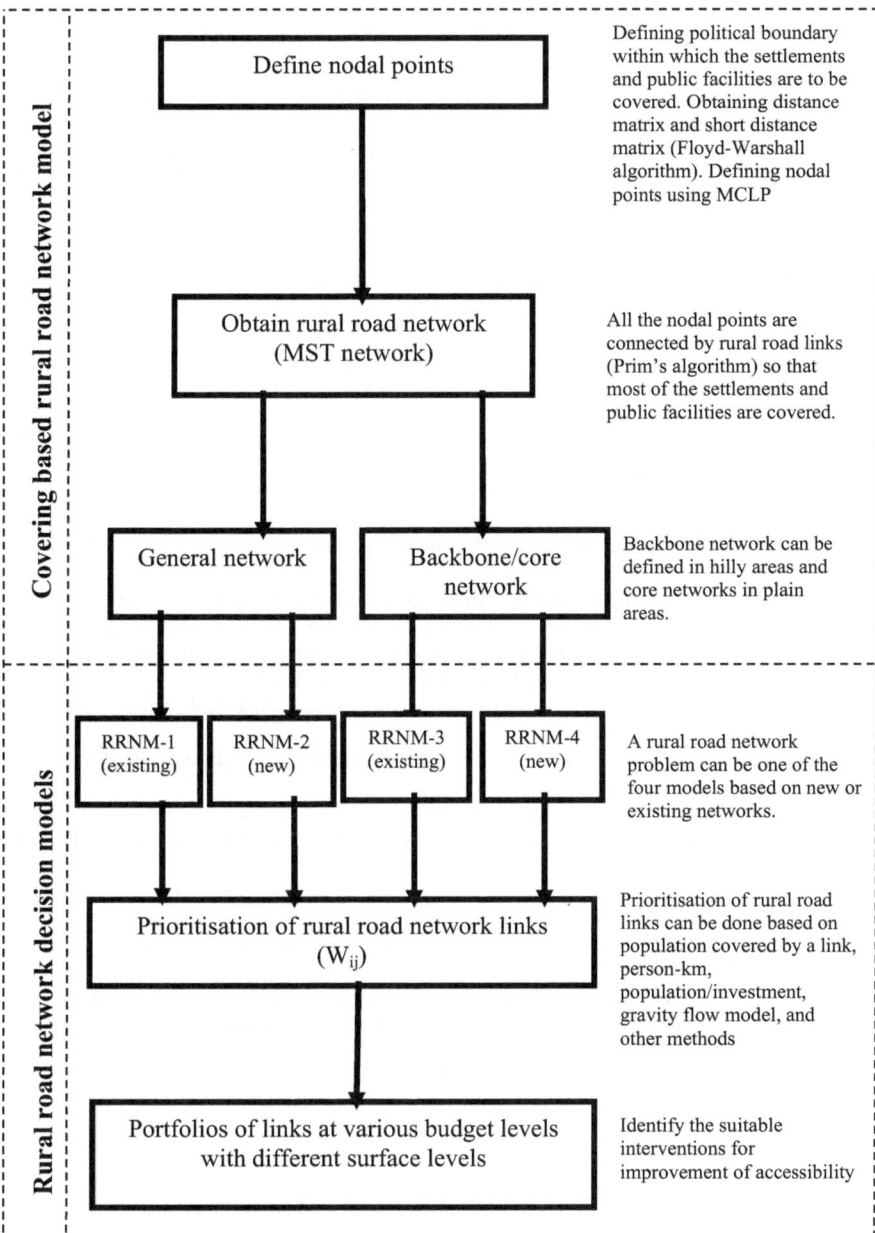

**Fig. 6.11** Proposed rural road network planning process

# References

1. DoLIDAR, Local Infrastructure Development Policy. (Ministry of Local Development, Government of Nepal, Kathmandu, 2004)
2. M.S. Daskin, S.H. Owen, Location models in transportation. In *Handbook of Transportation Science*, ed. by R.W. Hall (Kluwer Academic Publishers, Norwell, MA, ch.10, 1999), pp. 311–360
3. S. Melkote, M.S. Daskin, An integrated model of facility location and transportation network design. Transp. Res. Part A **35**, 515–538 (2001)
4. S. Heng, Y. Hirobata, H. Nakanishi, An integrated model of rural road network design and multi public facility locations in developing countries, in *Conference on Infrastructure Planning Committee, Japan Society of Civil Engineers*, vol.34 (2006).
5  J.K. Shrestha, A. Benta, R.B. Lopes, N. Lopes, C. Ferreira, A numerical model for rural road network optimization in hilly terrains, in *First ECCOMAS Young Investigators Conference* (YIC2012) (University of Aveiro, Portugal, 2012)
6. R.L. Church, C. ReVelle, The maximal covering location problem, Papers of the Regional Science Association 32, 101–118 (1974)
7. R.W. Floyd, Algorithm 97: Shortest Path. Commun. ACM **5**(6), 345 (1962). https://doi.org/10.1145/367766.368168
8. R.C. Prim, Shortest connection networks and some generalizations. Bell Syst. Tech. J. **36**, 1389–1401 (1957)
9. S. Heng, Y. Hirobata, H. Nakanishi, An integrated model of rural infrastructure design in developing countries, in *Proceedings of the Eastern Society for Transportation Studies* (2007)
10. T. Airey, G. Taylor, (1999). Prioritization procedure for improvement of very low-volume roads, Transportation Research Board of the National Academies **1652**.
11. K.E. Haynes, A.S. Fotheringham, *Gravity and Spatial Interaction Models, SAGE Publications, Beverly Hills* (Cal, USA, 1984)
12. W.S. Jung, F. Wang, H. Stanley, Gravity model in the Korean highway. EPL (Europhysics Letters), **81** (2007)
13. A. Kumar, H.T. Tilloston, A planning model for rural roads in India, in *Proceedings, Seminar on roads and road transport in rural areas* (Central Road Research Institute, New Delhi, India, 1985)
14. A.K. Makarachi, H.T. Tillotson, Road planning in rural areas of developing countries. Eur. J. Oper. Res. **53**, 279–287 (1991)
15. A. Kumar, P. Kumar, User friendly model for planning rural road. Transp. Res. Rec. **1652**, 31–39 (1999)
16. C.B. Shrestha, Developing a computer-aided methodology for district road network planning and prioritization in Nepal. Int. J. Transp. Manag. **1**(3), 157–174 (2003)

# Chapter 7
# A Multi-objective Analysis of Rural Road Networks

## 7.1 Introduction

A variety of goals can be employed to define or enhance rural road networks, as discussed in previous chapters. Development and network extension of rural roads are frequently associated with social and economic issues. Because of this, choices made with a single goal in mind could be irrational and poorly justified. It becomes a multi-objective strategy because several objectives must be addressed simultaneously. The rural road network problem is addressed in this chapter utilizing a multi-objective approach.

The optimal solution in single-objective techniques can be determined by ordering all possible alternatives depending on the objective function value. The concept of Pareto-optimality replaces the concept of optimal solution in multi-objective techniques. The multi-objective analysis finds the preferred solution—the one the DM deems interesting—instead of pursuing the best option.

Following are some fundamental ideas regarding multi-objective models. Next, a rural road network model will be subjected to a multi-objective strategy for connection selection and upgrading. Interactive and non-interactive approaches are the two available methods for solving multi-objective problems [1]. This investigation employed a non-interactive method of analysis. However, because the DM's function is critical and is dependent on the local context and plan, the DM must be properly informed throughout problem analysis.

## 7.2 Multi-objective Integer Programming: Basic Concepts

Several objectives are limited by a set of constraints in a general multi-objective integer programming problem. The problem can be expressed as follows:

(MOP)

J. Shrestha, *Rural Road Development in Developing Countries*,
SpringerBriefs in Applied Sciences and Technology,
https://doi.org/10.1007/978-981-96-2012-8_7

$$\text{max } Z_1 = f_1(x) \tag{7.1}$$

$$\text{max } Z_k = f_k(x) \tag{7.2}$$

Subject to:

$$x \in X \tag{7.3}$$

The number of objectives is $K$. $X \subset \mathbb{R}^n$ denotes the non-convex set of feasible solutions defined by the set of functional constraints, $x \geq 0$ and $x_j$ integer for $j \in J \subseteq \{1, 2, ..., n\}$. $X$ is assumed compact (closed and bounded) and non-empty.

A solution $\bar{x} \in X$ is efficient for the MOP if and only if there is no $x \in X$ such that $f_i(x) \geq f_i(\bar{x})$ for all $i \in \{1, 2, ..., k\}$ and $f_i(x) > f_i(\bar{x})$ for at least one $i$.

Let $Z \subset \mathbb{R}^k$ be the image of the feasible region $X$ in the objective functions space. A point $\bar{z} \in Z$ is called non-dominated if it corresponds to an efficient solution $\bar{x} \in X$. The phrases efficient, non-dominated, and Pareto-optimum are sometimes used interchangeably [1].

A non-dominated point (solution) $\bar{z} \in Z$ is said to be unsupported if it is dominated by a convex combination (not belonging to $Z$) of other non-dominated points (belonging to $Z$). Unsupported non-dominated solutions may exist because the feasible region is non-convex. As a result, unlike in multi-objective linear programming, the set of non-dominated solutions in MOP cannot be achieved completely by adjusting the value $\lambda$ on the weighted sum of the objective functions:
(MOP$_\lambda$)

$$\text{max } \sum_{i=1}^{k} \lambda_i f_i(x) \tag{7.4}$$

Subject to:

$$x \in X \tag{7.5}$$

where $\lambda \in \Lambda = \left\{ \lambda \in \mathbb{R}^k : \lambda_i > 0 \; \forall i, \; \sum_{i=1}^{k} \lambda_i = 1 \right\}$.

Even if the entire parameterization of $\lambda$ is attempted, unsupported non-dominated solutions are impossible to achieve. One solution is to add more constraints to MOP$_\lambda$, setting boundaries on the objective function values [2]:
(MOP$_{\lambda,\alpha}$).

$$\text{max } \sum_{i=1}^{k} \lambda_i f_i(x) \tag{7.6}$$

Subject to:

$$x \in X \tag{7.7}$$

$$Z_i \geq \alpha_i \; i = 1, \; 2, \; ..., \; k \tag{7.8}$$

where $\lambda \in \Lambda$ and $\alpha \in \mathbb{R}^k$.

Every non-dominated solution obtained by $\text{MOP}_\lambda$ is unique, and there is always an $\alpha$ such that $\text{MOP}_\lambda$ returns a specific non-dominated solution. Thus, $\text{MOP}_{\lambda,\alpha}$ allows one to determine the entire set of non-dominated MOP solutions. This method is known as weighted sum, and it can be applied with any parameterization of $\lambda$ to find solutions.

The optimal values are found by maximizing each objective function independently, and they are obtained by taking the greatest values of the objective functions over the set of efficient solutions $E$. In most cases, the resulting ideal point is not achievable; if it were, there would be no conflict and the solution would be ideal.

## 7.3 Objectives for Rural Road Network Problems

Many fields, like road network design, have dealt with multi-objective problems. However, there are still a few works underway in the area. Furthermore, works addressing the rural road network case are rarer. In terms of objectives, works in the literature can be classified as follows:

- Cost minimization [3–5].
- Equity [6–8].
- Robustness [9, 10].
- Accessibility maximization [6].
- Connectivity maximization [11].
- Minimization of travel distance [3, 9].
- Minimization of property expropriation [3].
- Minimization of carbon monoxide emissions [12].
- Other relevant objectives (e.g., route efficiency) [10].

The minimization of property expropriation and the minimization of travel distance were considered by Friesz and Harker [3] and [5] explored simultaneous minimization targets for user and building costs. [7] investigated two types of equity goals: vertical (across population centers within the same area) and horizontal (across all population centers). A paradigm that integrates fair aims and accessibility is offered by [6]. When allocating a specific budget for various prospective road projects, [10] established a methodology for designing rural road networks with two goals in mind: maximizing all-weather road connectivity between communities and maximizing route efficiency. In addition to trip time minimization as an efficiency goal, [9] included a robustness target.

The standard method for designing a road network is to identify the least expensive option that can manage a given set of traffic patterns [4]. But it ignores other socioeconomic factors like accessibility and the connection between traffic patterns and road conditions [6]. Nonetheless, the majority of studies have consistently prioritized minimum-cost goals, and one of the most important goals for rural roadway design issues can be reaching the minimum-cost target. When it comes to rural road network issues, construction and transit costs are the bare minimum [13, 14] . Network operation costs, however, might be a useful indicator of rural roadway optimization for upgrading rural roads. When road links are used, it measures the time, difficulty, distance, and costs.

An accessibility-maximization strategy was developed by [6] for the long-term development of interurban road networks. Accessibility is a critical component in rural population quality of life and regional development potential, this issue is particularly acute in rural locations. However, in steep areas, all settlements (people) cannot be connected. However, most villages may be reached in a decent amount of time. The covering-based rural road network model addresses this problem. The concept is covered in detail in Chap. 5. Maximizing the amount of inhabitants that can be included in a region is a social problem. This is an additional crucial goal to take into account when creating and modernizing rural road networks. In mountainous areas, measuring accessibility in terms of spatial measurement in a design is less important. It is more important to assess whether a road link can reach the region's communities and population. As a result, population covered is the best indication for maximizing accessibility in Nepal's hilly regions. The greater the population covered, the greater the accessibility.

While the aforementioned efforts are noteworthy, the objectives for rural road systems in remote regions may vary. Cost and accessibility criteria may be more relevant in this setting. As a result, in this study, two objectives are examined for the rural road optimization problem: minimizing user operation costs and increasing the population served by a network. In the first, an efficiency aim is discussed, whereas in the second, a social goal is highlighted.

Roads in rural areas are usually earthen, and they may require improvement to gravel or asphalt surfaces. Chapter 6 discusses the three types of road surfaces that are taken into consideration in this study: asphalt, gravel, and earthen.

When limited funds are available, a multi-objective analysis is undertaken to expand understanding of the rural road network problem.

## 7.4  Multi-objective Rural Road Model

When two objectives are examined for decision-making in this study, the model becomes a bi-objective problem, which is a subset of the multi-objective optimization problem (MOP) when two objectives are considered ($k = 2$).

Lowering the cost of user operation is the main objective under consideration. When upgrading from lower surface levels (gravel or earthen) to upper surface levels (gravel or asphalt), users are more inclined to select links with lower operational expenses. Due to their potential for cost savings, links with high traffic volumes are more likely to be selected. Rural traffic volume, however, is frequently low and challenging to assess. Because of this, the length of the link (distance) and the operating cost per unit flow of traveling across a surface are typically used to calculate the operation cost.

Increasing the linked population is the second goal. The likelihood of a link being selected for upgrade to higher surface levels increases with the size of the population it serves (even when the link's length and operating costs are both higher). It is considered as a societal objective. The population served by the link $(i, j)$ is indicated by $P_{ij}$.

An objective function of the model can be built to take into account the running costs with weights assigned to the linkages. Then, as the first objective, Eq. (7.1) can be restated.

Minimize

$$Z_1 = \sum_{S=1}^{3} \sum_{(i,j) \in L, i<j} W_{ij} O_{ij}^s x_{ij}^s \tag{7.9}$$

The second objective is to maximize the population coverage provided by the selected linkages.

Maximize

$$Z_2 = \sum_{S=1}^{3} \sum_{(i,j) \in L, i<j} P_{ij} x_{ij}^s \tag{7.10}$$

Subject to:

$$\sum_{S=1}^{3} \sum_{(i,j) \in L, i<j} I_{ij}^s x_{ij}^s \leq B \tag{7.11}$$

$$\sum_{S=1}^{3} x_{ij}^s = 1 \forall (i,j) \in L, \ i<j, \ \forall s \in S \tag{7.12}$$

$$x_{ij}^s \in \{0, 1\} \forall (i,j) \in L, \ i<j, \ \forall s \in S \tag{7.13}$$

Equation (7.11) states that the construction/improvement cost is restricted to an investment budget. Since the graph is undirected, the amount of money spent on creating only one connection, $(i, j)$ or $(j, i)$, is referred to as link construction expenditure. Limitations (7.12) stipulate that a particular type of surface can only be used on one link. These restrictions also guarantee that each linkage is connected to a surface option.

The model can be adjusted to account for additional limitations in a special instance. For instance, there are two kinds of links in mountainous regions: branch links and backbone links (Sect. 6.7). In the model, the backbone linkages are considered limitations that are updated prior to the branch links. Prior to dealing with branch links, the backbone linkages should be prioritized. Then, to incorporate the backbone and branch concept into the model, additional constraints can be introduced to the preceding model.

$$\sum_{S=2}^{3} x_{ij}^{s} \geq x_{kl}^{s} \forall (i,j) \in L, \ (k,l) \in L, \ i < j, \ k < l, \ j \leq m, \ m < l \qquad (7.14)$$

$$x_{ij}^{s} \in \{0, 1\} \forall (i,j) \in L, \ i < j, \ \forall s \in S, \qquad (7.15)$$

$$x_{kl}^{s} \in \{0, 1\} \forall (k,l) \in L, \ k < l, s \in S \qquad (7.16)$$

Subsidiary link nodes are numbered after backbone link nodes (see Chap. 5). The node count of the backbone network is represented by "$m$." Consequently, Eq. 7.14 assumes that selecting all of the backbone links is a prerequisite for selecting the secondary branch links.

This means that the node numbering in the network needs to be changed (refer to Fig. 6.2, where the links that connect nodes 1 through 10 are called backbone links, while the remaining links are called branch links).

## 7.5   Application of the Model

The applicability of the multi-objective model is assessed using the same rural roadway system that was covered in Chap. 5.

As mentioned in Chap. 5, person-km is the test priority parameter. Similar way, additional parameters can be added to the study. Each link is given the same weight by using Table 6.4. The number of trips is not included in this study since flow data for every link in this network is not available. A practical cost optimization analysis can be conducted using the travel data. It might be challenging to estimate travel statistics for rural places, particularly those with steep terrain. As previously noted, the links' earthen, gravel, and asphalt surfaces have operating costs per unit flow of NRs 50.64, NRs 45.64, and NRs 36.79. An estimated NRs 5 million will be required for each kilometer of upgrading from earth surface level to gravel surface level, and an additional NRs 10 million will be required for each kilometer of upgrading from gravel surface level to asphalt surface level.

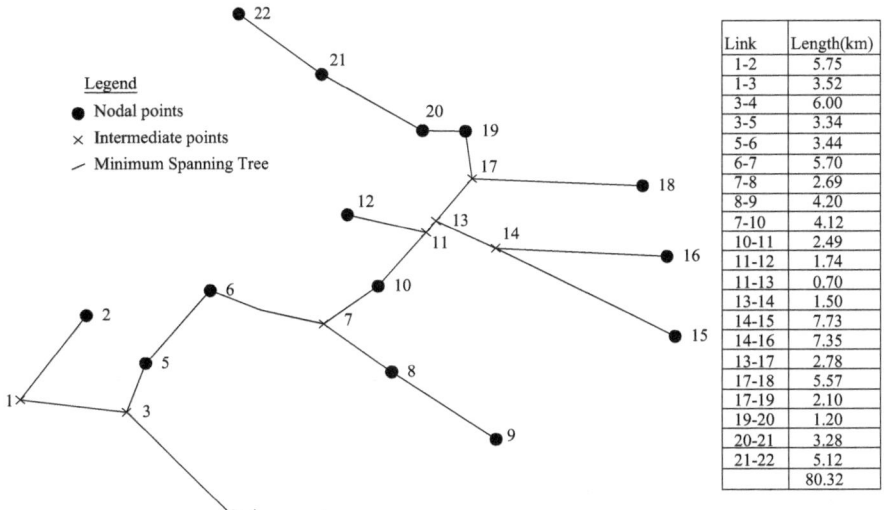

| Link | Length(km) |
|------|-----------|
| 1-2 | 5.75 |
| 1-3 | 3.52 |
| 3-4 | 6.00 |
| 3-5 | 3.34 |
| 5-6 | 3.44 |
| 6-7 | 5.70 |
| 7-8 | 2.69 |
| 8-9 | 4.20 |
| 7-10 | 4.12 |
| 10-11 | 2.49 |
| 11-12 | 1.74 |
| 11-13 | 0.70 |
| 13-14 | 1.50 |
| 14-15 | 7.73 |
| 14-16 | 7.35 |
| 13-17 | 2.78 |
| 17-18 | 5.57 |
| 17-19 | 2.10 |
| 19-20 | 1.20 |
| 20-21 | 3.28 |
| 21-22 | 5.12 |
| | 80.32 |

**Fig. 7.1**  Rural road network for application of model

## 7.5.1  Test Instance Data and Solutions

The weighted sum approach was used to solve the mathematical models utilizing MPL for Windows 4.2 as the modeling language and CPLEX 10.0's Mixed Integer Programming as the solver considering the rural road network as shown in Fig. 7.1. An analysis can be carried out under various budgetary limitations for the network.

The multi-objective model solution for the rural road network is produced for budget levels of NRs 400 million and NRs 600 million. The model offers a wide range of solutions. Any of the options may be chosen by the decision-maker. Nevertheless, they might consider some effective substitutes to achieve the goals. In Table 7.1, "a" denotes asphalt, "g" denotes gravel, and "e" denotes the earthen surface level of the road links.

## 7.5.2  Analysis of the Solutions

Non-dominated solution sets have been developed for the bi-objective models. The set of obtained solutions contains unique solutions. This indicates the decision-maker's preference, which is that all goals should be prioritized equally and that the closer an aim is to the ideal state, the better. Scattered throughout the Pareto-frontier are the Pareto-optimal solutions, which provide the decision-maker with a range of options.

Solutions shown in Table 7.1 are the non-dominated solutions for budget level NRs 400 millions. Similarly, the non-dominated solution for the budget level NRs

**Table 7.1** Non-dominated solutions for budget level NRs 400 millions

| Solutions | $Z_1$ | $Z_2$ | Links | | | | | | | | | | | | | | | | | | | | |
|---|---|---|---|---|---|---|---|---|---|---|---|---|---|---|---|---|---|---|---|---|---|---|---|
| | | | 1-2 | 1-3 | 3-4 | 3-5 | 5-6 | 6-7 | 7-8 | 7-10 | 8-9 | 10-11 | 11-12 | 11-13 | 13-14 | 13-17 | 14-15 | 14-16 | 17-18 | 17-19 | 19-20 | 20-21 | 21-22 |
| s1 | 73,060 | 396,592 | g | g | g | g | g | g | g | g | g | g | g | g | g | g | g | g | g | g | g | g | e |
| s3 | 70,690 | 395,292 | g | g | g | a | g | g | g | g | g | g | a | a | g | g | e | g | g | g | g | g | g |
| s5 | 69,981 | 393,867 | g | a | g | g | g | g | g | g | g | g | g | g | g | g | g | g | e | g | g | g | g |
| s8 | 69,936 | 390,884 | g | a | g | g | g | g | g | g | a | g | e | a | g | g | e | g | g | g | g | g | g |
| s10 | 69,740 | 388,939 | g | a | g | g | g | g | g | g | g | g | g | g | g | g | a | e | e | g | g | g | g |
| s20 | 68,717 | 378,985 | g | g | g | g | g | a | g | a | g | g | g | g | a | a | g | e | g | e | g | g | e |
| s22 | 66,001 | 377,658 | e | a | a | a | a | g | a | a | a | a | a | g | a | g | e | e | e | g | g | g | e |
| s28 | 61,005 | 372,275 | e | a | g | a | a | a | a | a | e | a | g | g | a | g | e | e | e | g | g | g | e |
| s48 | 60,302 | 361,601 | e | a | e | a | a | a | g | a | e | a | g | a | a | a | a | e | e | g | g | e | e |
| s61 | 60,208 | 357,193 | e | a | e | a | a | a | g | a | e | a | e | a | g | a | a | e | e | a | a | e | e |

600 millions is shown in Table 7.2. The Pareto-frontier for budget level NRs 400 millions is shown in Fig. 7.2. Pareto-frontier for the same network but for budget level NRs 600 millions was also obtained as shown in Fig. 7.3.

Figures 7.4 and 7.5 depict the decision alternatives for intervention in the rural road network with varying surface levels with budgets of NRs 400 million and NRs 600 million. The first budget level has 10 decision possibilities, and the second budget level has eight decision alternatives. Among them, four solutions which lies in the region where it has the lower operation cost and higher population coverage can be considered for comparison (Table 7.3).

We need to calculate the trade-offs between the two objective function values for the non-dominated set to assess the different possibilities for decision-making. Thus, we can evaluate the degree to which we must penalize the significance of one objective function to raise the value of the other. In this paradigm, every trade-off is positive; raising one objective function will raise the others. But in this case, we look for solutions with a lower value for the first goal, which is to minimize operational costs ($Z_1$), and a higher value for the second goal, which is to cover the population ($Z_2$). Moreover, we might be more interested in larger trade-offs, which imply that raising $Z_1$ is intended to result in the largest possible increase in $Z_2$ value. When we see solutions that have a higher $Z_1$ value but a lower $Z_2$ value, we may make smaller trades.

This method of identifying distinct solutions is made possible by the bi-objective model, which also enables the DM to evaluate objective values. The DM can analyze the trade-offs between the two objectives because they are familiar with a range of options that have a bigger population coverage and lower operating costs. Based on their preferences, the DM may choose a final solution that represents their judgment.

Table 7.4 indicates that at budget level NRs 600 million, the DM would be interested in the non-dominated solutions of the model that are located in the region with the lowest operation cost and higher population coverage space of the solution space.

The DM can then compare increases in population coverage with increases in operating costs, as indicated in Table 7.4.

s35 is the option with the lowest running cost among the others. Unfortunately, and contrary to what was intended, the solution only addresses a small segment of the population. s30, which covers 0.22% more people than s35 but has an operating cost increase of 0.48%, is the option with the second lowest operating cost. After s30, s29 follows with an operating cost that is 0.48% higher than s30. Decision-makers may not view this favorably if there is no discernible improvement in population coverage despite an increase in operating costs. The final option, s18, has a substantially higher rise in operation costs—4.44%—than the other options, but a less noteworthy increase in population coverage—just 0.37%. Solution s30, therefore, might represent an intriguing trade-off at the NRs 600 million budget level.

In addition to these analytical findings, the network's connectivity should be investigated in terms of market centers and the locations of public facilities. There is often a policy in place to link the development of rural road networks with market centers [13]. The majority of public facilities are often located in market centers. These aspects must also be addressed before making a decision.

**Table 7.2** Non-dominated solutions for budget level NRs 600 millions

| Solutions | $Z_1$ | $Z_2$ | Links | | | | | | | | | | | | | | | | | | | | |
|---|---|---|---|---|---|---|---|---|---|---|---|---|---|---|---|---|---|---|---|---|---|---|---|
| | | | 1–2 | 1–3 | 3–4 | 3–5 | 5–6 | 6–7 | 7–8 | 7–10 | 8–9 | 10–11 | 11–12 | 11–13 | 13–14 | 13–17 | 14–15 | 14–16 | 17–18 | 17–19 | 19–20 | 20–21 | 21–22 |
| s1 | 73,034 | 399,332 | g | g | g | g | g | g | g | g | g | g | g | g | g | g | g | g | g | g | g | g | g |
| s2 | 72,776 | 396,592 | g | g | g | g | g | g | g | g | g | g | g | g | g | a | g | g | g | g | g | g | e |
| s3 | 70,110 | 395,292 | g | a | g | g | g | g | g | g | g | g | g | g | g | g | e | g | g | g | g | g | g |
| s8 | 69,981 | 393,867 | g | a | g | g | g | g | g | g | g | g | g | g | g | g | g | g | e | g | g | g | g |
| s18 | 62,913 | 390,791 | e | g | a | a | a | a | a | a | a | a | g | a | a | a | e | a | g | a | a | g | g |
| s29 | 60,239 | 389,366 | e | a | g | a | a | a | g | a | g | a | g | a | g | a | g | g | e | a | a | g | g |
| s30 | 59,951 | 389,356 | g | a | g | a | a | a | g | a | a | a | a | a | g | a | g | a | e | a | a | e | e |
| s35 | 59,667 | 388,484 | g | a | g | a | a | a | g | a | e | a | g | a | g | a | a | a | e | a | g | a | g |

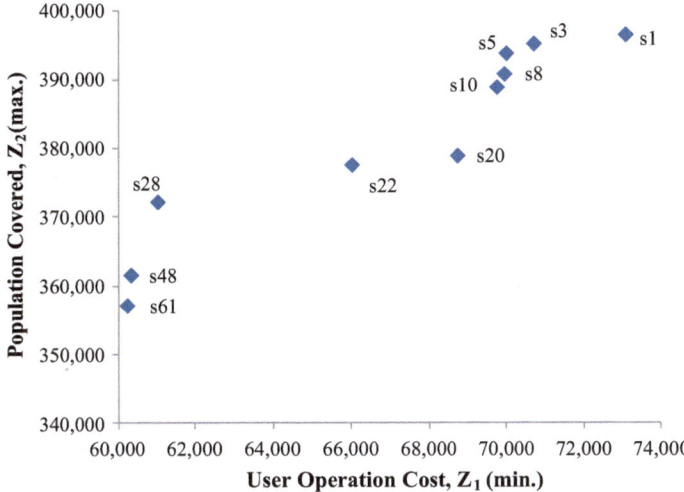

**Fig. 7.2**   Pareto-frontier for budget level NRs 400 millions

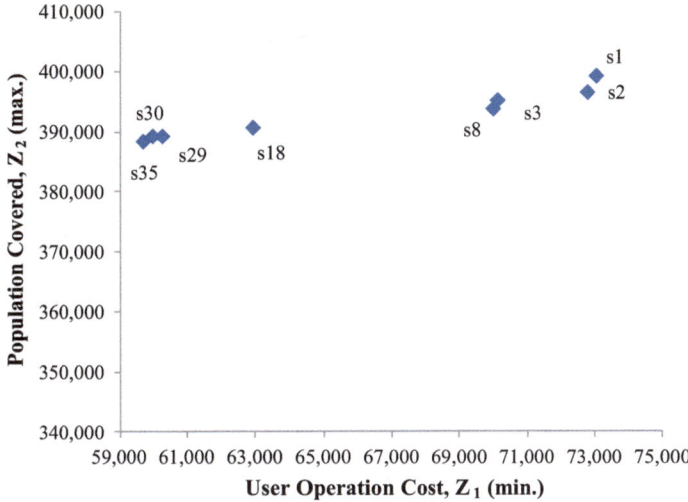

**Fig. 7.3**   Pareto-frontier for budget level NRs 600 millions

In this region, node 13 (Fig. 7.6) is home to a significant market center, but it is not a nodal point. It consists of essential public facilities such as bank, post office, police station, college, health clinic, and agricultural extension services. Consequently, a higher surface level will be added to the road links between node 13, the main market hub, and node 1, the main node. Similar public service stations can be found in another important market center around nodal point 16. Despite not being in this area, the market center is a major gateway to the northern part of the district. As a

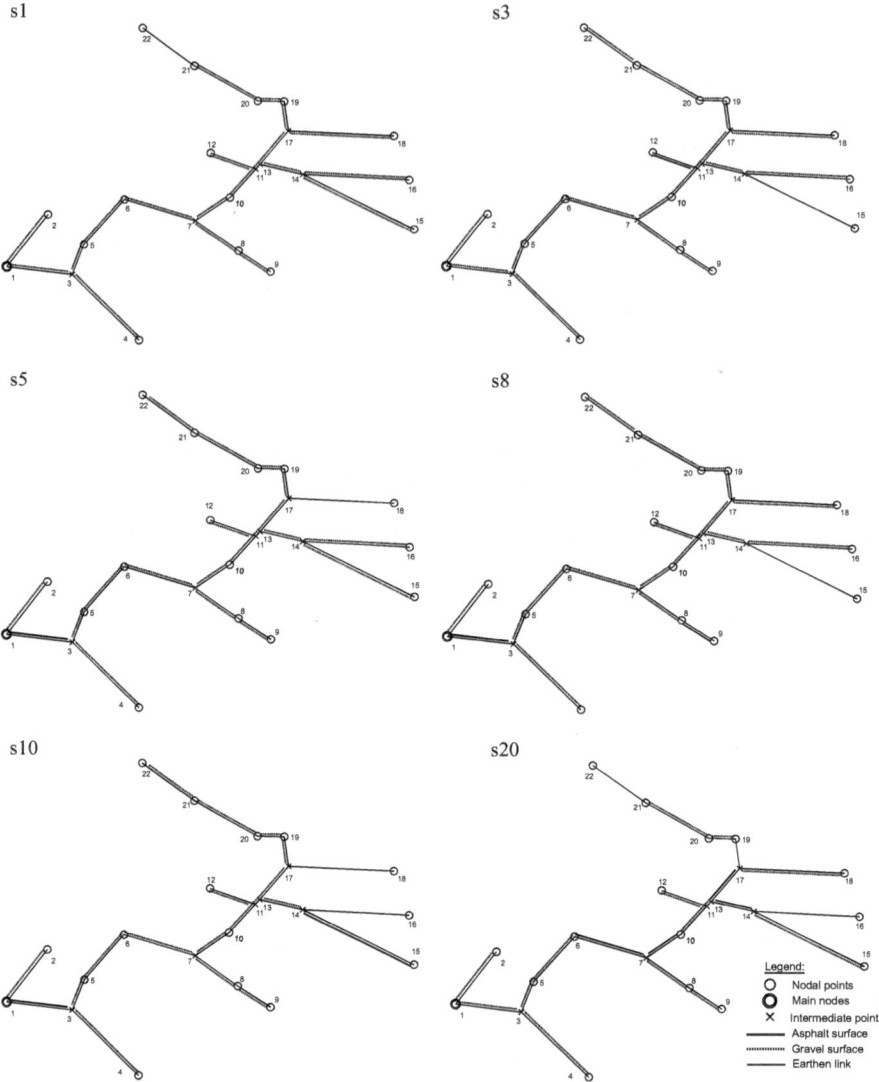

**Fig. 7.4**  Decision options and surface level of links for budget level NRs 400 millions

result, it is preferable to improve the surface levels of the links between nodes 13 and 16.

Of all the nodal sites, the communities surrounding node 12 have the densest population. When enhancing connections, this component can also be taken into account. This indicates that there is room for improvement of connection 11–12 to a higher surface level.

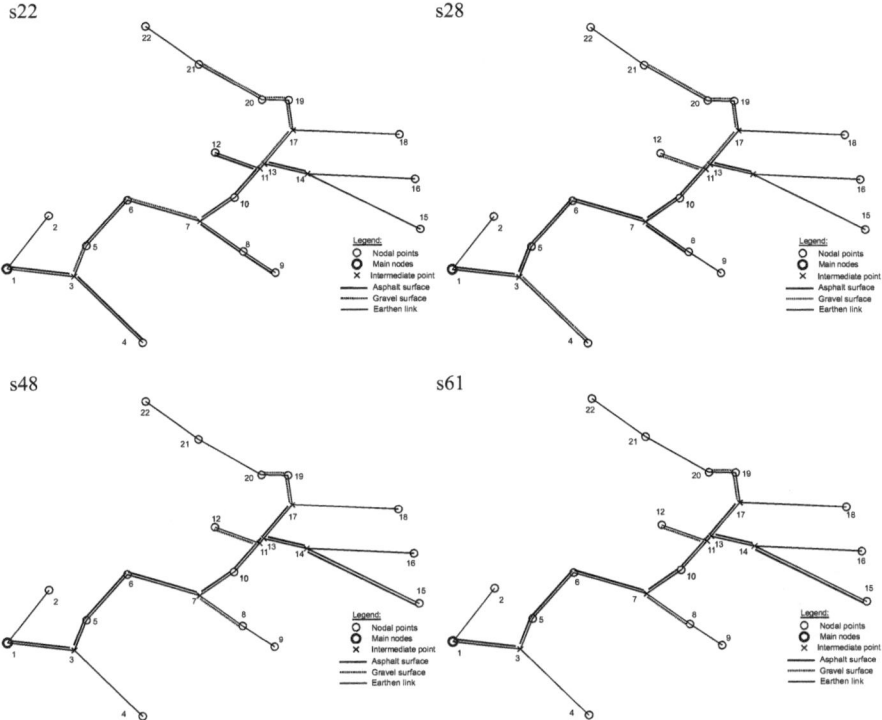

**Fig. 7.4** (continued)

Solutions s29 and s30 might be the best choices given the previously stated reasonable conditions. Both alternatives cover the densely populated areas and key marketplaces of the region. This analysis demonstrates that rather than a unique answer to the problem, it is possible to give the DM some interesting choices. Also, there are some opportunities in decisions to boost the value of one target without significantly increasing the value of the other.

Because of this, the DM may have a wide range of options that enable it to dramatically alter the population served by the rural road system without having to significantly raise operating costs. In this instance, there are numerous options available to the DM, enabling him to increase the value of expenses as well as the value of services. The DM should compare the improvements in the values of both objectives to those solutions that outperform the original. The DM, however, has the final say over which solution to implement. To do this, they must assess the model, weigh the pros and cons of each option, and consider any misconceptions they may have.

In a similar vein, it is possible to obtain and analyze multi-objective model solutions with branch linkages and a backbone, which would be appropriate for rural road networks situated in hilly regions.

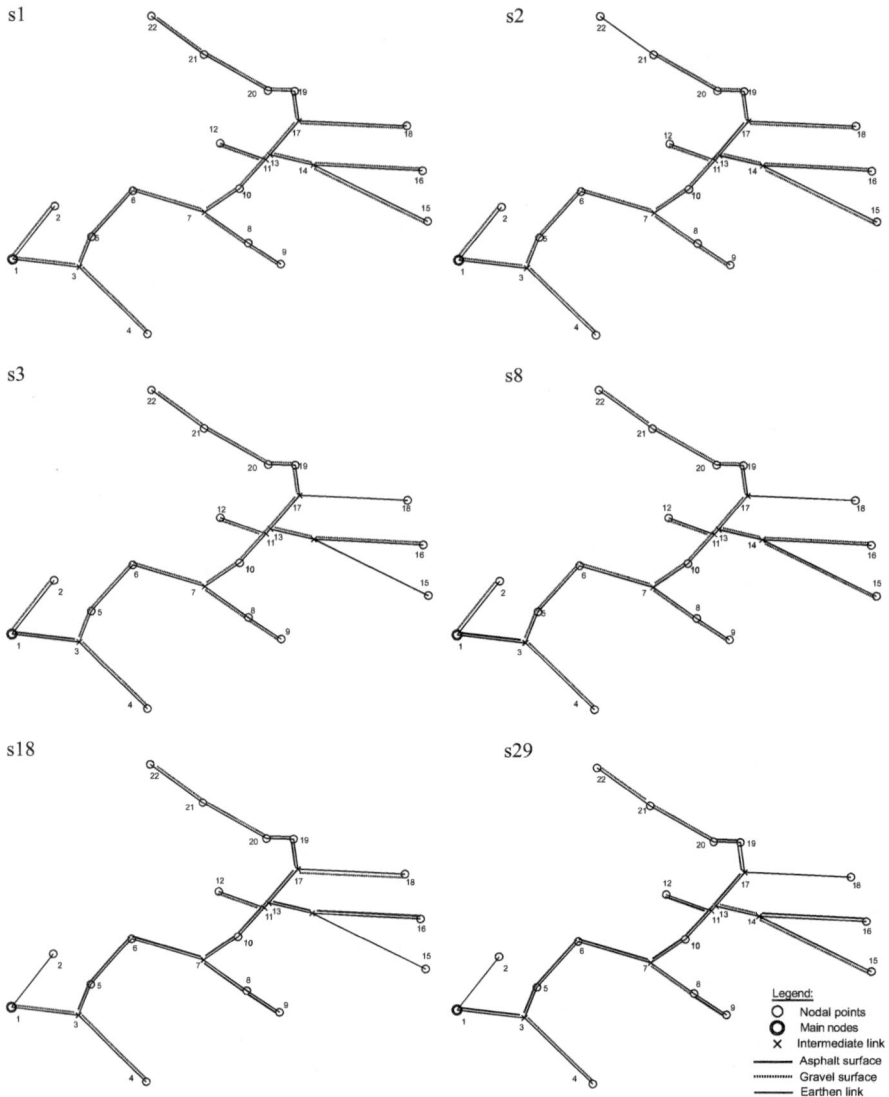

**Fig. 7.5** Decision options and surface level of links for budget level NRs 600 millions

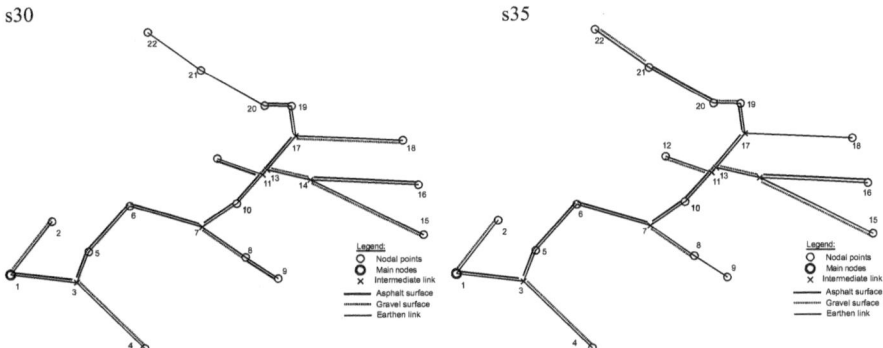

**Fig. 7.5** (continued)

## 7.6  Conclusions

This chapter addressed the rural road network problem as a multi-objective problem. A novel framework is suggested for broad rural road networks. The idea can also be applied to branch and backbone rural networks in hilly regions with minor adjustments. The rural road network that was being considered for Chap. 6 was used to test the model.

The proposed model minimizes operation expenses in the first objective and maximizes population coverage in the second, subject to a budgetary constraint.

With diverse optimal options dispersed over the Pareto-frontier, we can acquire all Pareto-optimal solutions that may be of interest to DM. This method was discovered to be suitable for effective study of rural road networks.

A flowchart (Fig. 7.7) summarizes the suggested planning procedure for rural road networks in rural areas using the multi-objective method. The first stage of the planning process (covering-based rural road network model) can be similar to that shown in Fig. 6.11.

The covering-based rural road network method was examined in the study. This method finds nodes that cover communities and public amenities and creates a basic road network connecting the nodes in an area to enhance physical access while making the most of limited resources. For the purpose of creating and upgrading new networks, the study also looked into a number of rural road network concepts. The applicability of the developed methodology was demonstrated when it was tested on actual road networks in the districts of Lamjung and Gorkha, Nepal, accounting for the placements of public facilities and communities. The models were determined to be realistic, easy to use, and useful in the setting of developing countries like Nepal.

**Table 7.3** Preferable solutions for budget level NRs 600 millions

| Solutions | $Z_1$ | $Z_2$ | Links | | | | | | | | | | | | | | | | | | | | |
|---|---|---|---|---|---|---|---|---|---|---|---|---|---|---|---|---|---|---|---|---|---|---|---|
| | | | 1–2 | 1–3 | 3–4 | 3–5 | 5–6 | 6–7 | 7–8 | 7–10 | 8–9 | 10–11 | 11–12 | 11–13 | 13–14 | 13–17 | 14–15 | 14–16 | 17–18 | 17–19 | 19–20 | 20–21 | 21–22 |
| s18 | 62,913 | 390,791 | e | g | a | a | a | a | a | a | a | a | g | a | a | a | e | a | g | a | a | g | g |
| s29 | 60,239 | 389,366 | e | a | g | a | a | a | g | a | g | a | g | a | g | a | g | g | e | a | a | g | g |
| s30 | 59,951 | 389,356 | g | a | g | a | a | a | g | a | a | a | a | a | g | a | g | a | g | a | a | e | e |
| s35 | 59,667 | 388,484 | g | a | g | a | a | a | g | a | e | a | g | a | g | a | a | a | e | g | g | a | g |

**Table 7.4** Comparisons of preferable solutions for budget level NRs 600 millions

| Solutions | $Z_1$ | $Z_2$ | Increase in % | |
|---|---|---|---|---|
| | | | Operation cost | Population coverage |
| s35 | 59,667 | 388,484 | | |
| s30 | 59,951 | 389,356 | 0.48 | 0.22 |
| s29 | 60,239 | 389,366 | 0.48 | 0.00 |
| s18 | 62,913 | 390,791 | 4.44 | 0.37 |

**Fig. 7.6** Map of the planning region

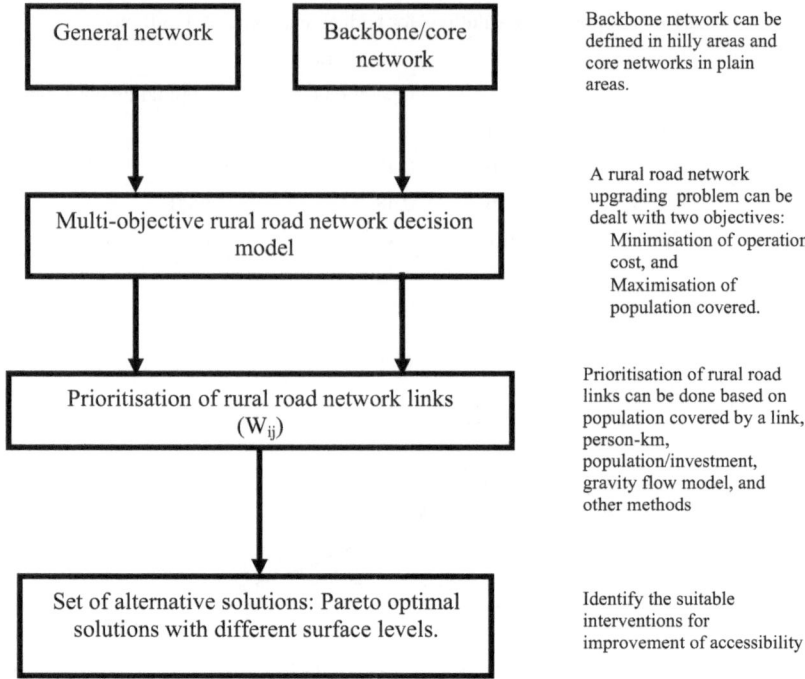

Backbone network can be defined in hilly areas and core networks in plain areas.

A rural road network upgrading problem can be dealt with two objectives:
  Minimisation of operation cost, and
  Maximisation of population covered.

Prioritisation of rural road links can be done based on population covered by a link, person-km, population/investment, gravity flow model, and other methods

Identify the suitable interventions for improvement of accessibility

**Fig. 7.7**   Proposed multi-objective rural road network planning process [14]

# References

1. M.J. Alves, J. Climaco, A review of interactive methods for multi-objective integer and mixed-integer programming (2006)
2. R.M. Soland, Multi-criteria optimization: a general characterization of efficient solutions. Decis. Sci. **10**(1), 26–38 (1979)
3. T.L. Friesz, G. Anandalingam, N.J. Mehta, K. Nam, S.J. Shah, R.L. Tobin, The multi-objective equilibrium network design problem revisited—a simulated annealing approach. Eur. J. Oper. Res. **65**(1), 44–57 (1993)
4. B.N. Janson, L.S. Buckels, B.E. Peterson, Network design programming of united-states highway improvements. J. Transp. Eng. **117**(4), 457–478 (1991)
5. G.H. Tzeng, S.H. Tsaur, Application of multiple criteria decision making for network improvement. J. Adv. Transp. **31**(1), 49–74 (1997)
6. A. Antunes, A. Seco, N. Pinto, An accessibility maximization approach to road network planning. Comput. Aided Civ. Infrastruct. Eng. **18**(3), 224–240 (2003)
7. C.M. Feng, J.Y.J. Wu, Highway investment planning model for equity issues. J. Urban Plann. Dev. **129**(3), 161–176 (2003)
8. Q. Meng, H. Yang, Benefit distribution and equity in road network design. Transp. Res., Part B: Methodol. **36**(1), 19–35 (2002)
9. S.V. Ukkusuri, T.V. Mathew, S.T. Waller, Robust transportation network design under demand uncertainty. Comput. Aided Civ. Infrastruct. Eng. **22**(1), 6–18 (2007)
10. H.K. Lo, Y.K. Tung, Network with degradable links: capacity analysis and design. Trans. Res. Part B: Method. (2003)

11. M.P. Scaparra, R.L. Church, A GRASP and path relinking heuristic for rural road network development. J. Heuristics **11**(1), 89–108 (2005)
12. G.E. Cantarella, A. Vitetta, The multi-criteria road network design problem in an urban area. Transportation **33**(6), 567–588 (2006)
13. A.K. Makarachi, H.T. Tillotson, Road planning in rural areas of developing countries. Eur. J. Oper. Res. **53**, 279–287 (1991)
14. A. Kumar, H.T. Tilloston, A planning model for rural roads in India, in *Proceedings, Seminar on roads and road transport in rural areas* (Central Road Research Institute, New Delhi, India, 1985)
15. C.B. Shrestha, Developing a computer-aided methodology for district road network planning and prioritization in Nepal. Int. J. Transp. Manag. **1**(3), 157–174 (2003)
16. J.K. Shrestha, *Rural engineering infrastructures design and public facility locations* (Universidade de Aveiro, Portugal, 2013)